Study Notes on 'A' Level Physics

DAVID LUCAS, M.A.

Senior Physics Master, The Bosworth School, Desford

Edward Arnold
London

SBN: 7131 2215 3

Photoset by The European Printing Corporation Limited Dublin Ireland

Printed in Great Britain by Cox & Wyman Ltd, London, Fakenham and Reading

PREFACE

It is common practice for students following a course in 'A' level Physics to spend a lot of their time making notes on the information given to them. This often results in the students missing part of what is being said, and also in errors in their notes.

These notes are offered, therefore, as one solution to these problems and also as suitable for general revision. They have evolved over a number of years of teaching 'A' level candidates and represent the notes that a conscientious pupil might make during the course. It has become the author's practice to issue these notes at the start of the course and so to make the course a discussion based on them.

The notes follow the pattern of a new type syllabus, and include a considerable amount of modern Physics at the expense of topics now being dropped by the examining boards. SI units and recommended symbols have been used throughout, although some mention has been made of other familiar units because it will probably be some years before they are finally abandoned.

The real is positive sign convention is used where necessary in the light section. The approach to electricity is similar to that of the 1966 A.S.E. report.

The order of the divisions has been chosen to lessen the mathematical demands at the start of a course, but they may be taken in any other order as desired.

No problems or worked examples are included as there are many excellent books containing these generally available.

The facts used in a book of this sort cannot be original, but have been gathered (with grateful thanks) from many sources.

The author would like to thank Mrs. G. Jessop for typing the manuscript and the publishers and their readers for all their help and guidance.

1969   D.J.L.

CONTENTS

vi   Contents

GEOMETRICAL OPTICS

1 REFLECTION AT A PLANE SURFACE

1.1 The laws of reflection

Reflection from a smooth polished surface follows a regular pattern and is called regular reflection, whereas reflection from a rough surface is called irregular or diffuse reflection.

Elementary work with a ray box or with pins will have shown that rays of light obey two simple laws when they are reflected at a smooth plane surface such as that of a mirror.

(*i*) The angle of incidence is equal to the angle of reflection.

(*ii*) The incident ray, the normal and the reflected ray all lie in the same plane.

If a point source of light is placed in front of a plane mirror and the path of various rays of light plotted, or constructed in accordance with these laws, it will be observed that all the reflected rays appear to come from one

point. It is therefore said that an image of the object is formed at this point.

The position of this image can be summarized in two simple rules.

(*i*) The line joining object and image is at right angles to the mirror.

(*ii*) The image is as far behind the mirror as the object is in front of it.

Geometrically these rules are equivalent to the laws stated above. They also explain why the image is perverted (laterally inverted).

1.2 The movement of a plane mirror

(*a*) *Movement perpendicular to the line joining object and image*

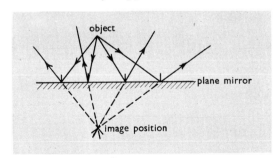

Fig. 1 The formation of an image

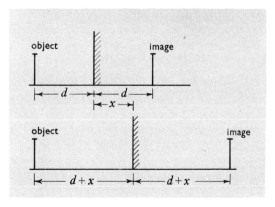

Fig. 2 Mirror moving back

When the mirror is moved a distance x, the distance from object to image increases by $2x$; i.e. the image moves twice as far and therefore twice as fast as the mirror.

(b) *Rotating mirror*

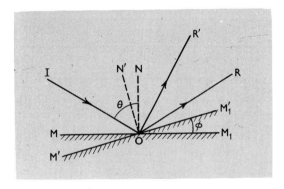

Fig. 3 Rotating mirror

The mirror MM_1 rotates through the angle ϕ to $M'M'_1$, the normal moving from ON to

ON′ and the reflected ray from OR to OR′. The incident ray does not move.

$$\hat{ION} = \hat{NOR} = \theta \qquad \text{(Laws of reflection)}$$

$$\hat{N'ON} = \hat{M'_1OM_1} = \phi \qquad \text{(Normal moves as mirror)}$$

$$\therefore \quad \hat{ION'} = \hat{ION} - \hat{NON'}$$
$$= \theta - \phi$$
$$= \hat{N'OR'} \qquad \text{(Laws of reflection)}$$

$$\hat{NOR'} = \hat{N'OR'} - \hat{N'ON}$$
$$= \theta - \phi - \phi$$
$$= \theta - 2\phi$$

$$\therefore \quad \hat{R'OR} = \hat{NOR} - \hat{NOR'}$$
$$= \theta - (\theta - 2\phi)$$
$$= 2\phi$$

Thus the reflected ray moves through twice the angle turned through by the mirror.

2 REFRACTION AT A PLANE SURFACE

2.1 The laws of refraction

(*i*) The incident ray, the normal and the refracted ray all lie in the same plane.

(*ii*) The ratio of the sine of the angle of incidence to the sine of the angle of refraction is a constant for all rays of the same wavelength passing from one given substance to another.

2.2 Refractive index

When light passes through a parallel-sided block the emergent ray is found to be parallel

to the incident ray so that the two angles marked α are equal.

The angles marked β are equal by geometry (alternate).

In order to specify fully the refractive index, the constant from the second law, it is necessary to specify the media concerned.

$$_A n_G = \frac{\sin \alpha}{\sin \beta} \qquad _G n_A = \frac{\sin \beta}{\sin \alpha}$$

$$\therefore \quad _A n_G = \frac{1}{_G n_A} \quad \text{or} \quad _A n_G \cdot _G n_A = 1$$

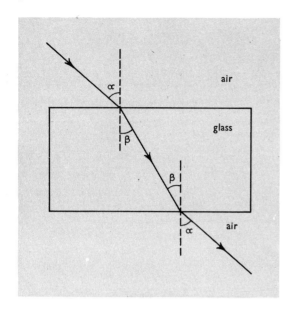

Fig. 4　Refraction through a rectangular glass block

In more elaborate cases, using two parallel layers of material in air, similar relationships hold.

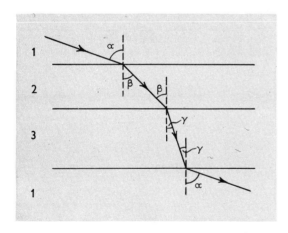

Fig. 5　Refraction through parallel layers

The angles α are found to be equal by experiment, the others by geometry.

$$_1n_2 = \frac{\sin \alpha}{\sin \beta}$$

$$_2n_3 = \frac{\sin \beta}{\sin \gamma}$$

$$_3n_1 = \frac{\sin \gamma}{\sin \alpha}$$

$$\therefore \quad _1n_2 \cdot {}_2n_3 \cdot {}_3n_1 = \frac{\sin \alpha}{\sin \beta} \cdot \frac{\sin \beta}{\sin \gamma} \cdot \frac{\sin \gamma}{\sin \alpha} = 1$$

or

$$_1n_2 \cdot {}_2n_3 = \frac{1}{_3n_1} = {}_1n_3$$

2.3　Absolute refractive index

This is the refractive index for light travelling from a vacuum into the material. As the absolute refractive index of air is 1·00029 at s.t.p. there is little difference between the absolute refractive index and the refractive index for light passing from air into the material as

$$_{\text{vac.}}n_{\text{med.}} = {}_{\text{vac.}}n_{\text{air.}} \cdot {}_{\text{air}}n_{\text{med.}}$$

2.4　Total internal reflection

When light is travelling into an optically rarer medium from an optically denser medium, the angle of refraction is greater than the angle of incidence and must therefore reach 90° first. At this position, the critical case, the light travels along the surface of separation after refraction. When the angle of incidence increases further, all the light is reflected. Partial reflection occurs before this.

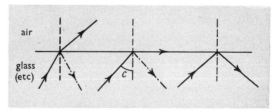

Fig. 6　Total internal reflection

At the critical case, where the angle of incidence is called the critical angle,

$$_G n_A = \frac{\sin c}{\sin 90} = \sin c$$

$$_A n_G = \frac{1}{_G n_A} = \frac{1}{\sin c}$$

2.5   The measurement of refractive index

(a)   Ray tracing

Using any shape of block, rectangular for simplicity, the angles of incidence and refraction are measured and hence the refractive index is calculated.

(b)   Real and apparent depth methods

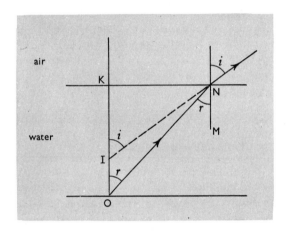

Fig. 7   Real and apparent depth

Light from O is refracted at N appearing to come from I.

$$\overset{\wedge}{\text{KIN}} = i \quad \text{corresponding}$$

$$\overset{\wedge}{\text{KON}} = r \quad \text{alternate}$$

$$_A n_W = \frac{\sin i}{\sin r} = \frac{\text{KN/IN}}{\text{KN/ON}}$$

$$= \frac{\text{NO}}{\text{NI}}$$

and if the eye is over the object so restricting the rays used to a narrow pencil NO = KO and NI = KI

$$\therefore \quad _A n_W = \frac{\text{KO}}{\text{KI}} \quad \text{or} \quad \frac{\text{Real depth}}{\text{Apparent depth}}$$

(*i*)   Vertical Position. Place a block of transparent material on a sheet of paper, then using a travelling microscope, focus on (a) paper, (b) paper with block in way, (c) top of block

$$_A n_M = \frac{\text{(a)} - \text{(c)}}{\text{(b)} - \text{(c)}}$$

(*ii*)   Horizontal Position

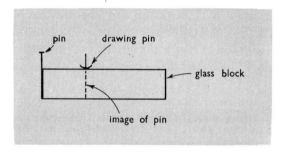

Fig.   8   Real   and   apparent   depth-horizontal measurement

Obtain a position of no-parallax between the drawing pin and the image of the pin.

(*c*)   *For a liquid—using a concave mirror*
See §4.8

(*d*)   *For a liquid—using a convex lens*
See §5.6

(*e*)   *For a liquid, air cell method*

The liquid is placed in a cubical cell and the air cell (a glass–air–glass sandwich) is rotated until the pins on the far side to the eye are on the point of disappearing, i.e. refraction is changing to total internal reflection.

The air cell is then turned to the symmetrical position and the angle turned through (2θ) is measured

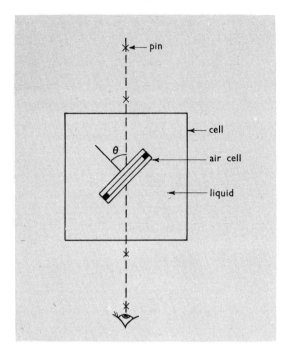

Fig. 9   The air cell

c is critical angle (glass to air)

$$c = \phi \text{ alternate}$$

$$_L n_G = \frac{\sin \theta}{\sin \phi} = \frac{\sin \theta}{\sin c}$$

$$_A n_G = \frac{1}{\sin c}$$

$$\therefore \quad _L n_G = \sin \theta \cdot _A n_G$$

$$\therefore \quad \frac{_L n_G}{_A n_G} = \sin \theta$$

$$= _L n_G \cdot _G n_A = _L n_A$$

$$\therefore \quad _A n_L = \frac{1}{\sin \theta}$$

Thus θ is equal to the critical angle for light going from liquid to air.

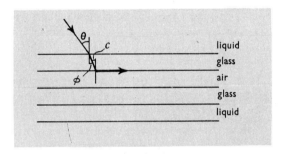

Fig. 10   Theoretical diagram for the air cell

3   PRISMS

3.1   General relationships

A = refracting angle of prism

D = angle of deviation

Angle marked at P $= A$ by angle sum of a quadrilateral and angle on a straight line.

$A = r_1 + r_2$ (exterior angle of $\triangle$)

$$D = (i_1 - r_1) + (i_2 - r_2) \quad \text{(exterior angle of } \triangle\text{)}$$

$$= i_1 + i_2 - (r_1 + r_2)$$

$$= i_1 + i_2 - A$$

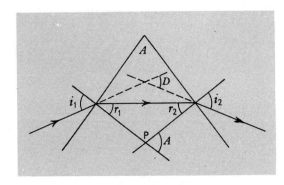

Fig. 11   Refraction through a prism

To proceed further a problem must be worked from first principles unless it is one of two special cases, a small angle prism at near normal incidence, or a large angle prism at minimum deviation.

3.2   A small angle prism at near normal incidence

Using the above diagram:

     A is small by definition

     i_1 is small by definition

     r_1 is small by Snell's law

     r_2 is small as it is $< A \, (A = r_1 + r_2)$

     i_2 is small by Snell's law

     (Small means that $\sin \theta = \theta$ in radians).

so

$$n = \frac{\sin i_1}{\sin r_1} = \frac{i_1}{r_1}$$

$$\therefore \quad i_1 = nr_1 \quad \text{similarly} \quad i_2 = nr_2$$

$$\therefore \quad D = i_1 + i_2 - r_1 - r_2$$

$$= nr_1 + nr_2 - r_1 - r_2$$

$$= (n-1)(r_1 + r_2)$$

$$= (n-1)A \quad \text{N.B. Independent of } i_1.$$

3.3   The dispersion produced in this case

The angular dispersion between two colours is the difference in deviation for these colours

e.g.

$$D_R = (n_R - 1)A \quad D_B = (n_B - 1)A$$

$$\therefore \quad \text{dispersion} = D_B - D_R = (n_B - n_R)A$$

where D_B = deviation for blue light

          n_B = refractive index for blue light, etc.

3.4   The actual deviation in a special case

$$A = 8° \qquad\qquad n = 1\cdot 5$$

i	0°	10°	20°	30°	40°	50°	60°	70°	80°
D	4°3′	4°3′	4°13′	4°39′	5°24′	6°36′	8°36′	11°50′	17°9′

3.5   Two thin prisms combined

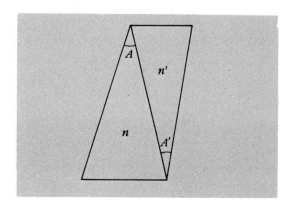

Fig. 12   Two small angle prisms combined

	Prism 1	Prism 2
Let the refracting angles be	A	A'
Let the refractive index for colour$_1$ be	n_1	n_1'
Let the refractive index for colour$_2$ be	n_2	n_2'
Let the mean refractive index be	n	n'

Then the total mean deviation

$$= (n-1)A - (n'-1)A'$$

(— as the two prisms work against each other)

And the total dispersion

$$= (n_2 - n_1)A - (n_2' - n_1')A'.$$

So by suitably choosing the angles, either no deviation, or no dispersion can be obtained.

For no deviation $(n-1)A - (n'-1)A' = 0$

$$\frac{A'}{A} = \frac{n-1}{n'-1}$$

For no dispersion $(n_2-n_1)A$
$$- (n_2'-n_1')A' = 0$$

$$\frac{A'}{A} = \frac{n_2-n_1}{n_2'-n_1'}$$

3.6 The deviation produced by a large angle prism

Experiment or calculation based on the laws of refraction applied to each face of the prism in turn, yield graphs as shown.

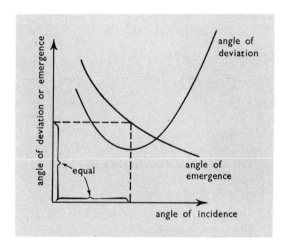

Fig. 13 Deviation in a large angle prism

Two main points should be observed:

(i) The deviation shows a minimum value.

(ii) When this minimum deviation occurs the angle of incidence is equal to the angle of emergence, i.e. the light passes symmetrically through the prism.

3.7 The minimum deviation formula

Using the property of symmetry·

$$D = 2(i-r)$$

$$A = 2r$$

$$\therefore \quad r = \frac{A}{2} \quad i = \frac{D+A}{2}$$

$$n = \frac{\sin i}{\sin r} = \frac{\sin \frac{1}{2}(D+A)}{\sin \frac{1}{2}A}$$

N.B. The '$\frac{1}{2}$'s' and the 'sine's' cannot be cancelled.

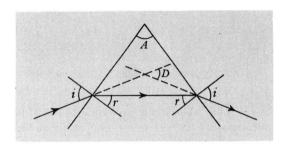

Fig. 14 Minimum deviation

3.8 The spectrometer

This instrument enables the previous formula to be applied with great accuracy.

The instrument consists basically of a collimator (to produce a parallel beam of light), a telescope (to receive the parallel beam of light) and a table (to hold the prism, etc.)

3.9 Preliminary adjustments

(i) Looking through the telescope towards a uniformly illuminated area, the eyepiece is adjusted so that the crosswires are clearly in focus.

(ii) Looking through the telescope towards a distant object the telescope is focussed so that there is no parallax between the crosswires and the image.

(iii) Looking straight through telescope and collimator at any light source, the collimator is adjusted so that a clear image of the slit is seen, and there is no parallax between the crosswires and this image.

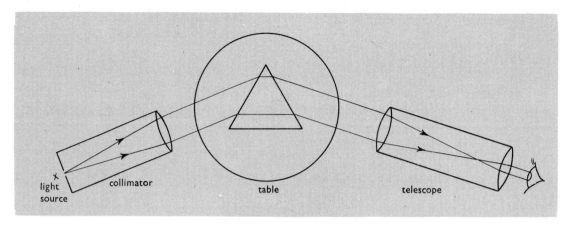

Fig. 15 Spectrometer

(*iv*) The prism table is levelled by viewing reflections of the slit in the two faces of the prism by dividing the beam from the collimator as in Fig. 16 and centring the images on the crosswires.

3.10 Measurements

(*a*) *The angle of the prism*

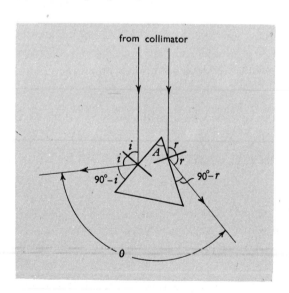

Fig. 16 Measurement of the angle of a prism

θ is measured (using both verniers), using any light source.

$$2i + \theta + 2r = 360° \quad \text{(Angle sum at a point)}$$
$$90 - i + A + 90 - r = \theta$$

multiply through by 2

$$\therefore \quad 360 - 2i - 2r + 2A = 2\theta$$

add, $\qquad\qquad\qquad \theta + 2A = 2\theta$

$$\theta = 2A$$

(*b*) *The angle of minimum deviation*

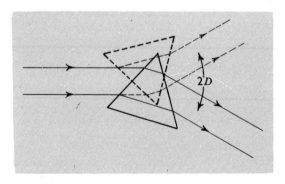

Fig. 17 Measurement of the angle of minimum deviation

The telescope is set in such a position that the particular line in use moves up to the crosswires and back as the table is steadily turned. A source of light giving a line spectrum must be used.

3.11 Spectra

These may be observed with

(*i*) a single prism (when they are impure due to overlapping);

(*ii*) a single prism and a lens (when they are much purer due to the focussing action of the lens);

(*iii*) a spectrometer (which is the purest as parallel light falls on to the prism);

(*iv*) a direct vision spectroscope using a series of prisms or a diffraction grating.

The spectra obtained depend upon the source but the following terms are used.

(*a*) *Emission spectra*

The light comes directly from the source.

(*i*) Continuous spectra—one colour merges into the next (e.g. from a filament electric light bulb)

(*ii*) Line spectra—a series of discrete lines (e.g. from a sodium lamp)

(*b*) *Absorption spectra*

The light which would give a continuous emission spectrum is passed through a medium which absorbs certain colours so that the result is a continuous spectrum crossed by dark lines.

4 SPHERICAL MIRRORS

4.1 Geometrical terms

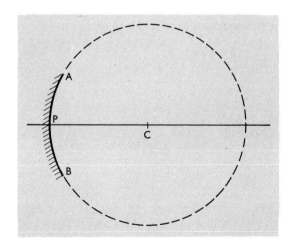

Fig. 18 Geometry of a spherical mirror

Using the idea that a spherical mirror is one that is part of a sphere, the following can be defined:

(*i*) C is the centre of curvature.

(*ii*) P is the pole.

(*iii*) PC is the radius of curvature.

(*iv*) The line through P and C is the principal axis.

(*v*) AB is the aperture.

4.2 The action on parallel rays of light

Treat each point on the mirror as plane, then the rays near to the pole are brought to a focus at F but those striking the mirror further out cross the principal axis nearer to the pole. The envelope of the reflected rays is called a caustic curve.

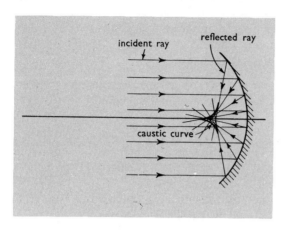

Fig. 19   Caustic curve

Thus to obtain a true focus one must restrict attention to rays near to the principal axis, or use a paraboloidal mirror.

4.3   The relation between focal length and radius of curvature

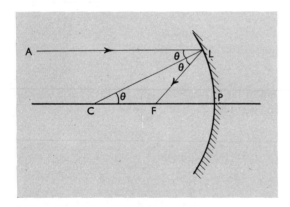

Fig. 20   Focal length and radius of curvature

Consider a ray of light parallel to the principal axis striking the mirror at L (near to P) so that it is reflected through the principal focus F.

The normal at L passes through C.

Treating the mirror at L as plane, apply the laws of reflection, making

$$A\hat{L}C = C\hat{L}F$$

but $A\hat{L}C = L\hat{C}F$ (alternate)

so $C\hat{L}F = L\hat{C}F$

$\therefore$ $\triangle$LCF is isosceles

so $LF = FC$

and as L is near to P

$$LF \simeq PF$$

so

$$PF = FC \quad \text{or} \quad PF = \tfrac{1}{2}PC$$

$\therefore$ the focal length equals half the radius of curvature.

4.4   The graphical construction of images

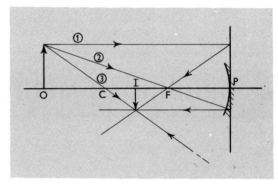

Fig. 21   Ray diagram–concave mirror

The mirror must be represented by a straight line having the properties of the mirror for accurate work.

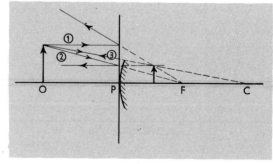

Fig. 22   Ray diagram–convex mirror

Ray 1, travelling parallel to the principal axis is reflected through F.

Ray 2, travelling through F is reflected parallel to the principal axis.

Ray 3, travelling through C is reflected back along its own path.

All the rays meet at the tip of the image.

4.5　The mirror formula

(a)　Concave mirror

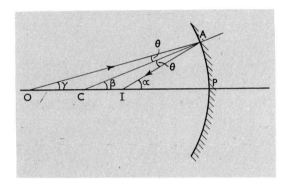

Fig. 23　Concave mirror formula

Imagine a luminous point object at O, giving rise to a point image at I.

As the laws of reflection must be obeyed at A

$$O\hat{A}C = I\hat{A}C \quad \text{(AC is the normal at A)}$$

Consider the angles marked

$$
\begin{array}{ll}
\text{in } \Delta IAC & \alpha = \beta + \theta \\
\text{in } \Delta CAO & \beta = \theta + \gamma
\end{array}
\right\} \text{ exterior } \angle \text{ of } \Delta
$$

$$\therefore \quad \theta = \alpha - \beta = \beta - \gamma$$

$$\therefore \quad \alpha + \gamma = 2\beta$$

but as A is near to P

$$\alpha = \frac{AP}{IP} \quad \beta = \frac{AP}{PC} \quad \gamma = \frac{AP}{PO}$$

$$\therefore \quad \frac{AP}{IP} + \frac{AP}{OP} = \frac{2AP}{PC}$$

Let　　　$IP = v \quad OP = u \quad PC = r$

$$\frac{1}{u} + \frac{1}{v} = \frac{2}{r}$$

$$= \frac{1}{f} \quad \text{(proved)}$$

Now no complications arise in this formula until $u < f$, then v becomes negative. But under these conditions the image is virtual, so a negative value of u, v, r or f is associated with a virtual quantity. This gives the sign convention called real is positive. (All distances measured to real objects, images or foci count positive, all distances measured to virtual objects, images or foci count negative. The radius of curvature takes the same sign as the focal length).

(b)　Convex mirror

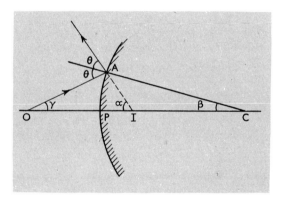

Fig. 24　Convex mirror formula

With the same ideas and notation as before

$$\text{in } \Delta ACI \quad \alpha = \beta + \theta$$

$$\text{in } \Delta ACO \quad \theta = \beta + \gamma$$

$$\therefore \quad \theta = \alpha - \beta = \beta + \gamma$$

$$\therefore \quad \gamma - \alpha = -2\beta$$

$$\therefore \quad \frac{AP}{PO} - \frac{AP}{PI} = -2\frac{AP}{PC}$$

but　　　$PO = u \quad PI = -v \quad PC = -r$

$$\therefore \quad \frac{1}{u} + \frac{1}{v} = \frac{2}{r} = \frac{1}{f}$$

4.6 Linear magnification

This is defined as

$$\frac{\text{height of image}}{\text{height of object}}$$

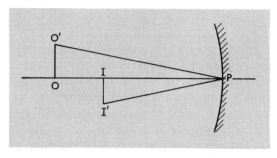

Fig. 25 Magnification in a concave mirror

As the mirror element at P is perpendicular to the axis

$$O'\hat{P}O = I\hat{P}I' \quad \text{laws of reflection}$$

$$P\hat{O}O' = P\hat{I}I' \quad \text{right angle}$$

$$\therefore \quad \Delta OPO' \text{ and } \Delta IPI' \text{ are similar}$$

so

$$\text{magnification} = \frac{\text{height of image}}{\text{height of object}} = \frac{II'}{OO'}$$

$$= \frac{IP}{OP} = \frac{\text{image distance}}{\text{object distance}} = \frac{v}{u}$$

4.7 The experimental determination of the focal length and radius of curvature of spherical mirrors

(a) Concave and convex

(i) The Spherometer. This consists of three fixed legs at the corners of an equilateral triangle, with a moveable leg at the centre. The instrument is first placed on a level surface, the zero read and then placed on the curved surface and the movement of the centre leg recorded.

$$AE^2 = AB^2 - BE^2$$

$$= 4l^2 - l^2 = 3l^2$$

$$AE = \sqrt{3}l$$

but $$AD = \tfrac{2}{3}AE = \frac{2\sqrt{3}l}{3} = \frac{2l}{\sqrt{3}}$$

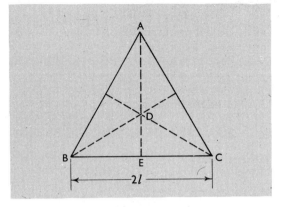

Fig. 26 Spherometer — plan view

Imagine B moved round about the centre D until ABD are all in one plane.

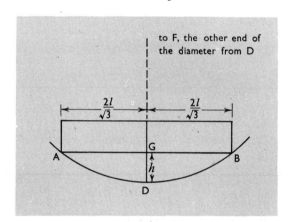

Fig. 27 Spherometer — side view

$$AG \cdot GB = GD \cdot GF$$

$$\left(\frac{2l}{\sqrt{3}}\right)^2 = h(2r - h)$$

$$\frac{4l^2}{3} = 2rh - h^2$$

$$r = \frac{2l^2}{3h} + \frac{h}{2}$$

$$= \frac{4l^2 + 3h^2}{6h}$$

where r = radius of curvature of surface

N.B. Other formulae exist where l is differently defined.

(ii) The Conjugate Point Method. With an illuminated object and screen (concave only) or with pins and no parallax (concave and convex) a series of values for the position of the object and image can be found.

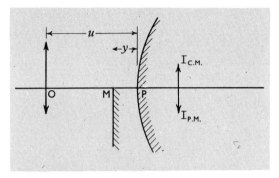

Fig. 28 Conjugate point method using a plane mirror

N.B. For virtual images it is not easy to view the search pin over the mirror and get it into the no-parallax position, so either the centre part of the mirror's silvering is removed and the search pin viewed through the hole, or a plane mirror is used.

The two images are adjusted to be in no-parallax, then

$$\text{image distance (IP)} = \text{IM} - \text{PM}$$
$$= \text{OM} - \text{PM}$$
$$= u - y - y$$
$$= u - 2y$$

The values of u and v can be interchanged, effectively increasing the range of observations.

If graphs of the results are plotted the focal length may be read off.

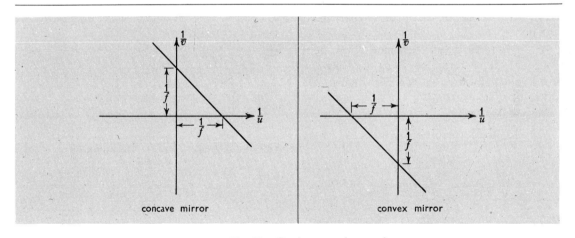

concave mirror   convex mirror

Fig. 29 Conjugate point graphs

(b) *Concave only*

(iii) Distant Object Method. The image of a distant object is formed at the principal focus and if it is thrown on a screen, the distance from mirror to screen is easily measured.

(iv) Self-Conjugate Point Method

A clamp pin is moved until it is in no parallax with its own image. Thus the light must be reflected back along its own path and for this to occur the object and the image must be at the centre of curvature. The radius of curvature is measured directly (as the mean of three readings).

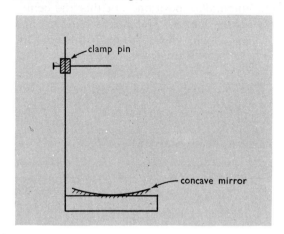

Fig. 30   Self-conjugate point method

4.8   The refractive index of a liquid using a concave mirror

The technique is the same as that of the last method above, the self-conjugate point is found for the mirror alone and when filled with liquid.

If C' is to be a self-conjugate point, light from it must be refracted so as to strike the mirror normally (i.e. along CM), then it will return along its own path.

Thus at the surface

$$C'\hat{S}N' = i$$
$$M\hat{S}N = r$$

Also $S\hat{C}L = r$ (corresponding) and $S\hat{C}'L = i$ (alternate)

so
$$\frac{\sin i}{\sin r} = \frac{SL/SC'}{SL/SC} = \frac{SC}{SC'} = n$$

now S is usually near to L

so
$$n = \frac{CL}{C'L}$$

or providing only a thin layer of liquid is used

$$n = \frac{CP}{C'P}$$

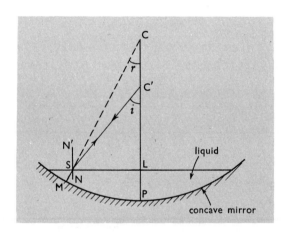

Fig. 31   Refractive index of a liquid using a concave mirror

5 THIN LENSES

The meaning of the following terms should already be known: centres of curvature, radii of curvature, principal axis, optical centre, principal focus and focal length.

5.1 The full formula for a thin lens

Sometimes the action of a lens is compared to that of a stack of prism sections of gradually changing angle. An extension of this enables a full formula for a lens to be developed.

A point object at O gives rise to an image at I. Consider the path of a ray as shown, refracted at the edge of the lens. C_1 and C_2 are the centres of curvature and the two sloping dotted lines the tangents to the lens surface at the edge.

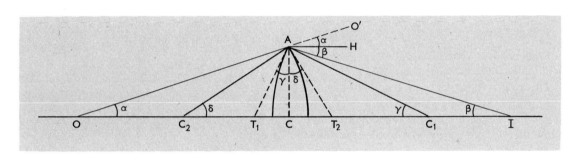

Fig. 32 Full formula for a thin lens

Consider the angles as marked along the axis, $O'\hat{A}H = A\hat{O}C = \alpha$ corresponding, $H\hat{A}I = A\hat{I}C = \beta$ alternate − where AH is parallel to OI.

$$T_1\hat{A}C = 90 - C\hat{A}C_1 \text{ (radius } \perp \text{ tangent)}$$
$$= 90 - (90 - \gamma) \text{ (angle sum of } \triangle)$$
$$= \gamma$$

likewise $T_2\hat{A}C = \delta$

Now imagine prism element at edge of lens.

Deviation $= O'\hat{A}I = \alpha + \beta$

Angle of prism $= T_1\hat{A}T_2 = \gamma + \delta$

and as prism is of small angle

deviation $= (n-1)$ refracting angle

$$\therefore \quad \alpha + \beta = (n-1)(\gamma + \delta)$$

but as the angles are small

$$\alpha = \frac{AC}{CO} \quad \beta = \frac{AC}{CI} \quad \gamma = \frac{AC}{CC_1} \quad \delta = \frac{AC}{CC_2}$$

$$\therefore \quad \frac{AC}{CO} + \frac{AC}{CI} = (n-1)\left(\frac{AC}{CC_1} + \frac{AC}{CC_2}\right)$$

$$\therefore \quad \frac{1}{CO} + \frac{1}{CI} = (n-1)\left(\frac{1}{CC_1} + \frac{1}{CC_2}\right)$$

and with usual notation

$$\frac{1}{u}+\frac{1}{v}=(n-1)\left(\frac{1}{r_1}+\frac{1}{r_2}\right)=\frac{1}{f}$$

$$(\text{as } v=f \text{ when } u=\infty)$$

If the lens is not in air

$$\frac{1}{u}+\frac{1}{v}=\left(\frac{n_L-n_S}{n_S}\right)\left(\frac{1}{r_1}+\frac{1}{r_2}\right)$$

where n_L = absolute refractive index of lens
n_S = absolute refractive index of surroundings

N.B. If $n=1.5$ for an equiconvex lens in air,

$$\frac{1}{f}=(1.5-1)\left(\frac{1}{r}+\frac{1}{r}\right)$$

$$f=r \text{ (cf. spherical mirror)}$$

but this is only a special case; there is generally no simple connection between f and r.

The real is positive sign convention gives r a positive sign when the surface is convex to the less dense medium, and a negative sign when the surface is concave to the less dense medium. Thus in Fig. 33 if $n_2 > n_1$ r is positive and if $n_2 < n_1$ r is negative.

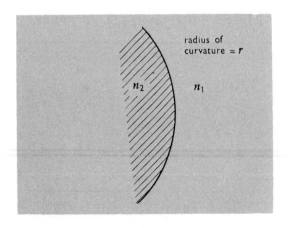

Fig. 33   Sign convention applied to radius of curvature

5.2   Two thin lenses in contact

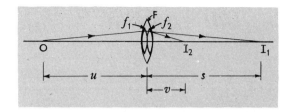

Fig. 34   Two thin lenses in contact

Consider two thin lenses of focal lengths f_1 and f_2 and one lens of focal length F, which does the same job as the other two.

For the first lens only, the real object at O gives a real image at I_1.

$$\frac{1}{u}+\frac{1}{s}=\frac{1}{f_1} \qquad (1)$$

For the second lens only, the virtual object at I_1 gives a real image at I_2

$$-\frac{1}{s}+\frac{1}{v}=\frac{1}{f_2} \qquad (2)$$

N.B. As the lenses are thin, s can be measured to either centre.

For the equivalent lens, the real object at O gives a real image at I_2

$$\frac{1}{u}+\frac{1}{v}=\frac{1}{F} \qquad (3)$$

add $(1)+(2)$

$$\frac{1}{u}+\frac{1}{v}=\frac{1}{f_1}+\frac{1}{f_2}$$

from (3)

$$=\frac{1}{F}$$

$$\therefore \frac{1}{f_1}+\frac{1}{f_2}=\frac{1}{F}$$

5.3   The graphical construction of images

A straight line to represent the lens must be used for accurate work.

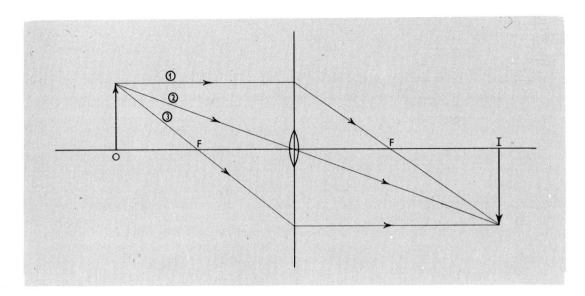

Fig. 35   The graphical construction of images for
a convex lens

Ray 1, travelling parallel to the principal axis
is refracted through F.

Ray 2, travelling through the optical centre is
not refracted.

Ray 3, travelling through F, is refracted
parallel to the principal axis.

All the rays meet at the tip of the image.

5.4   Methods of measuring the focal length of a convex (converging) lens

(a)   *Distant object method*

The image of a distant object will be formed
on a screen placed at the principal focus and
the focal length may be measured directly.

(b)   *Conjugate point method*

Using an illuminated object and screen
and/or pins and no parallax a series of
values for u and v can be found. Note that
only pins allow virtual images to be found.

The results can be plotted in three ways and
values of f can be found in each case.

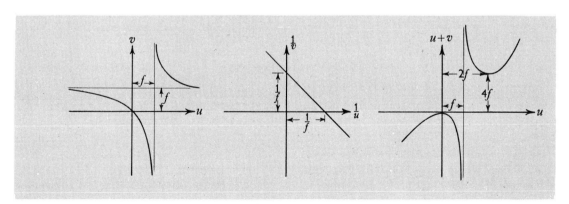

Fig. 36   Graphs for the convex lens

Note in the last case there is a minimum value of $(v+u)$. This can be predicted by calculus.

As

$$\frac{1}{u}+\frac{1}{v}=\frac{1}{f}$$

$$\frac{1}{v}=\frac{1}{f}-\frac{1}{u}=\frac{u-f}{uf}$$

$$\therefore \qquad v=\frac{uf}{u-f}$$

$$u+v=\frac{uf}{u-f}+u$$

$$=\frac{uf+u^2-uf}{u-f}$$

$$=\frac{u^2}{u-f}$$

$$\therefore \quad \frac{d(u+v)}{du}=\frac{2u(u-f)-u^2.1}{(u-f)^2}$$

$$=0 \quad \text{when} \quad 2u^2-2uf-u^2=0$$

$$\text{or} \quad u^2-2uf=0$$

$$\text{so} \quad u=0 \quad \text{or} \quad u=2f$$

when $u=2f$ $u+v=4f$ and this is the minimum value. In this position $u=v=2f$, so the position is symmetrical and object and image are the same size.

(Note: This means an object and real image are never nearer than $4f$).

(c) Self-conjugate point method

N.B. In use the lens is on the mirror. In the diagram it is drawn above the mirror for clarity.

The pin is only in no parallax with its own image if the light returns along its own path, i.e. strikes the mirror normally. By considering the return path of the light the pin must be at the principal focus.

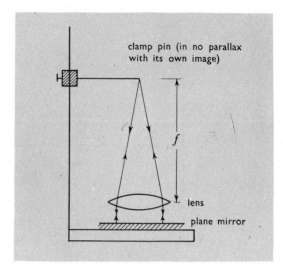

Fig. 37   Self-conjugate point method for a convex lens

(d) Displacement method

In general there are two positions of a lens between an object and screen, so that the image is clearly focussed on the screen.

This experiment can be conveniently carried out on an optical bench. A distance piece must be used to find the distance between the illuminated object and the screen. The distance moved by the lens from one position to the other is exactly the same as that moved by the lens holder.

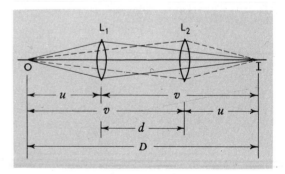

Fig. 38   The displacement method for a convex lens

The two positions of the lens must be symmetrical (to satisfy the lens equation and the fixed value of $u+v$)

Then $\quad D = d + 2u \quad \therefore \quad u = \dfrac{D-d}{2}$

$$v = u + d = \dfrac{D+d}{2}$$

substitute into $\quad \dfrac{1}{u} + \dfrac{1}{v} = \dfrac{1}{f}$

$$\dfrac{2}{D-d} + \dfrac{2}{D+d} = \dfrac{1}{f}$$

$$\dfrac{2D + 2d + 2D - 2d}{D^2 - d^2} = \dfrac{1}{f}$$

$$\dfrac{1}{f} = \dfrac{4D}{D^2 - d^2}$$

$$f = \dfrac{D^2 - d^2}{4D}$$

(e) Newton's formula

If the principal foci are located by (say) method c, but the lens is inaccessible (in a tube perhaps) the focal length may be found from the values of the distance from the object to the nearest focus and from the image to the nearest focus.

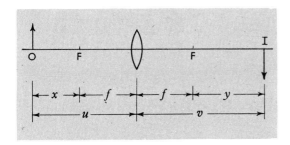

Fig. 39   Newton's formula

$$u = f + x \qquad v = f + y$$

substitute into $\quad \dfrac{1}{u} + \dfrac{1}{v} = \dfrac{1}{f}$

$$\dfrac{1}{f+x} + \dfrac{1}{f+y} = \dfrac{1}{f}$$

$$\therefore \quad f(f+y) + f(f+x) = (f+x)(f+y)$$

$$\therefore \quad f^2 + fy + f^2 + fx = f^2 + fx + fy + xy$$

$$\therefore \quad f^2 = xy$$

(f) Magnification method

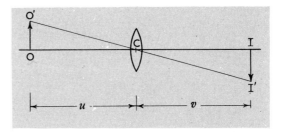

Fig. 40   Magnification formula

Considering the undeviated ray through the optical centre of the lens

Δ's OCO' and ICI' are similar.

so

$$\dfrac{II'}{OO'} = \dfrac{IC}{OC} \quad \text{(ratio of similar sides)}$$

but

$$\dfrac{II'}{OO'} = \dfrac{\text{size of image}}{\text{size of object}}$$

$$= \text{magnification}$$

$$= \dfrac{IC}{OC} = \dfrac{v}{u}$$

so magnification

$$= \dfrac{v}{u}$$

as

$$\dfrac{1}{u} + \dfrac{1}{v} = \dfrac{1}{f}$$

$$\dfrac{v}{u} + 1 = \dfrac{v}{f}$$

$$m = \dfrac{v}{f} - 1$$

And if magnification and v are measured directly with an illuminated object and screen, f may be found graphically.

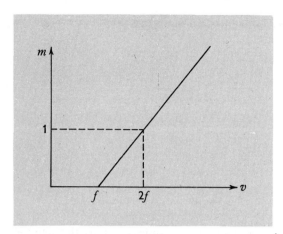

Fig. 41 Graph for magnification produced by a convex lens

(g) *Using a subsidiary lens of shorter focal length*

(Especially suitable for a long focal length lens)

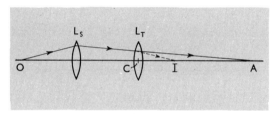

Fig. 42 Use of subsidiary lens

An image is obtained with the subsidiary lens (L$_s$) alone. The position (A) is recorded. The lens under test (L$_T$) is inserted and the new image position (I) is recorded.

Then A acts as a virtual object giving a real image at I

$$\left.\begin{array}{l} u = -AC \\ v = IC \end{array}\right\} \quad \text{for} \quad L_T$$

$$\therefore \quad -\frac{1}{AC} + \frac{1}{IC} = \frac{1}{f}$$

5.5 Methods of measuring the focal length of a concave (diverging) lens

(a) *Conjugate foci*

The technique of no parallax with pins must be used. The search pin must be viewed above the lens.

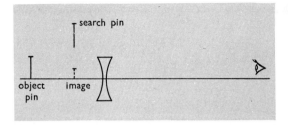

Fig. 43 Self-conjugate point method for a concave lens

A graph of $1/v$ against $1/u$ is plotted

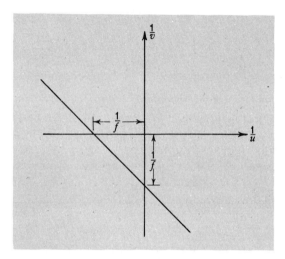

Fig. 44 Graph for the concave lens

As the image and the search pin are separated a plane mirror can be used:

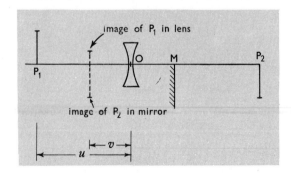

Fig. 45 Use of subsidiary plane mirror with a concave lens

The two images of the respective pins are obtained in no parallax

$$u = P_1O \qquad v = -(P_2M - MO)$$

(b)   Using a subsidiary convex lens in contact

A suitable convex lens is chosen so that the combination is still convex. The focal length of this combination and of the convex lens are measured by any suitable method. The formula for lenses in contact is then used to find the focal length of the concave lens.

(c)   Using a subsidiary convex lens not in contact

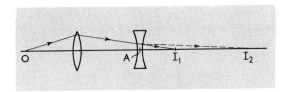

Fig. 46   Use of subsidiary convex lens with a concave lens

I_1 is the image produced by the convex lens alone, I_2 with both lenses, so I_1 acts as a virtual object for the concave lens, giving a real image at I_2

$$\therefore \quad u = -AI_1 \qquad v = +AI_2$$

A series of readings may be obtained by moving one lens relative to the other.

(d)   Using a subsidiary convex lens not in contact and a plane mirror

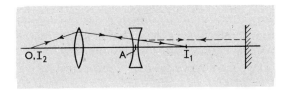

Fig. 47   Use of subsidiary convex lens and a plane mirror with a concave lens

The image is formed at the same point as the object and the focal length of the concave lens $= AI_1$ where I_1 is the image position with the convex lens alone.

(e)   Using a subsidiary concave mirror

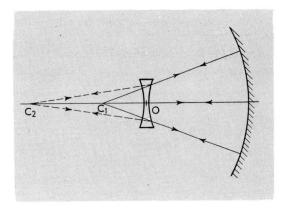

Fig. 48   Use of a subsidiary concave mirror with a concave lens

C_1 is the self-conjugate point without the lens.
C_2 is the self-conjugate point with the lens.
C_1 acts as a virtual object giving a real image at C_2.

$$\therefore \quad u = -C_1O \qquad v = +C_2O$$

and a series of values of u and v may be used to find the focal length.

5.6   The determination of the refractive index of a liquid

Find the focal length of the convex lens using the plane-mirror method.

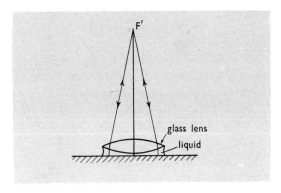

Fig. 49   Use of a convex lens to determine the refractive index of a liquid

Place some liquid beneath the lens, where it acts like a concave lens. The focal length of the combined lens is then found by the same method.

Using the formula for lenses in contact

$$\frac{1}{OF'} = \frac{1}{OF} + \frac{1}{f}$$

where f = focal length of liquid lens

and
$$\frac{1}{f} = (n-1)\left(\frac{1}{r} + \frac{1}{\infty}\right)$$

where n = refractive index of the liquid,

     r = radius of curvature of the upper liquid lens surface.

and therefore of the lower glass lens surface.

5.7   Boy's method for the radius of curvature of a lens

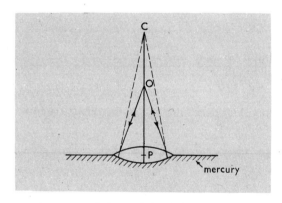

Fig. 50   Boy's method

The self-conjugate point for light reflected from the lower lens surface is found. Floating the lens on mercury makes this easier.

The light must strike the lower surface normally if it is reflected back, and so the rays in the glass:

    (i)   appear to have come from the centre of curvature of the lower surface,

    (ii)   would emerge from the lower surface with no further refraction

so the lens equation may be applied, O then being a real object giving a virtual image at C

$$\therefore \quad \frac{1}{OP} - \frac{1}{CP} = \frac{1}{f}$$

hence CP may be found.

5.8   The power of a lens

$$\text{Power} = \frac{1}{\text{Focal length}}$$

and power is in dioptres if the focal length is in metres. This is used mainly by opticians.

If two thin lenses are placed in contact their resultant focal length (F) has been shown to be given by

$$\frac{1}{F} = \frac{1}{f_1} + \frac{1}{f_2}$$

where f_1 and f_2 are the focal lengths of the separate lenses. So in terms of power

$$P = p_1 + p_2$$

where P = power of combined lenses
     p_1, p_2 = powers of separate lenses

5.9   The defects caused by a thin lens

In developing the theory of lenses certain restrictions were imposed (e.g. that the rays were near to the principal axis). If a full theory is worked out certain other effects occur, giving rise to defects in the image. Such defects are known as aberrations, and result in a point object giving an illuminated patch as an image, instead of another point.

Two main types of aberration occur:

(a)   Spherical aberration

If parallel rays fall on to the lens, those near to the axis converge to I, those near to the extremities to I', and intermediate ones between I and I'. The image is extended, and a circle of least confusion is obtained at the best position somewhere between I and I'.

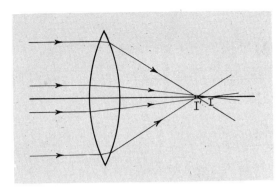

Fig. 51   Spherical aberration

The amount of aberration depends on the shape of the lens, and is least when the bending of the rays is shared equally between the two surfaces.

A plano-convex lens can be used to illustrate this.

In both cases the same total deviation occurs, but in (*i*) all the bending occurs at the second surface, while in (*ii*) it is shared equally. The resulting spherical aberration is much less in (*ii*).

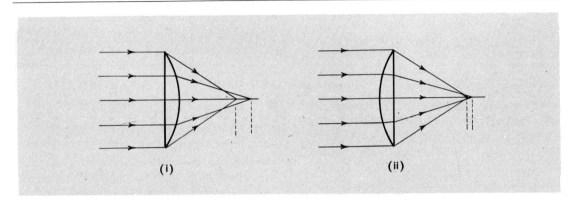

Fig. 52   Reduction of spherical aberration

(*b*)   *Chromatic aberration*

Considering the lens as a series of prisms, it follows that different colours will be deviated to different extents. This again causes a spread of the image, and also causes it to be falsely coloured.

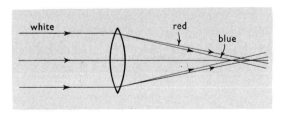

Fig. 53   Chromatic aberration

It can be partially corrected by using an achromatic doublet. This is two lenses in contact, made of different glasses, to produce deviation but no dispersion. Strictly it only corrects for two colours, the ends of the spectrum being overlapped. Some aberration still occurs.

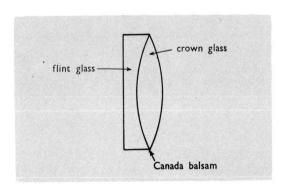

Fig. 54   An achromatic doublet

6 OPTICAL INSTRUMENTS

The human eye, its defects and their correction, the camera and the projector should have been discussed in elementary work.

6.1 The simple microscope

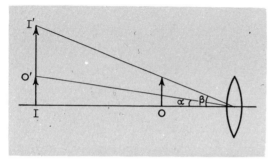

Fig. 55 The angular magnification of a simple microscope

A magnified image is formed at or beyond the near point of the eye. The term angular magnification is basically defined as follows:

$$\text{Angular magnification} = \frac{\text{Angle subtended at eye by image}}{\text{Angle subtended at eye by object}}$$

but assuming that the eye is near to the lens this is always one which makes it of no value. The definition breaks down because the object could not be seen if the lens were removed, so in this case

$$\text{Angular magnification} = \frac{\text{Angle subtended at eyepiece by image}}{\text{Angle subtended at eyepiece by object at near point}}$$

$$= \frac{\beta}{\alpha}$$

and if the image is at the near point also

$$= \frac{II'}{O'I} = \text{linear magnification}$$

6.2 The compound microscope

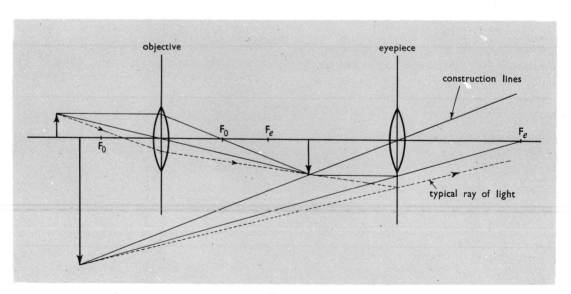

Fig. 56 The compound microscope

The objective is usually of very short focal length and the eyepiece of longer, but still short, focal length.

The difference between construction lines and rays of light must be carefully noted.

$$\text{Linear magnification} = \frac{\text{Size of final image}}{\text{Size of object}}$$

$$\text{Angular magnification} = \frac{\begin{array}{c}\text{Angle subtended at eyepiece by}\\ \text{final image}\end{array}}{\begin{array}{c}\text{Angle subtended at eyepiece by}\\ \text{object at near point}\end{array}}$$

If the final image is at the near point these are equal.

Problems are best solved in stages, applying the usual lens equations to each lens in turn.

Note that if the final image is at infinity the intermediate image subtends the same angle as the final image at the eyepiece, equal to the height of the intermediate image over the focal length of the eyepiece.

In actual microscopes the lenses are compound to minimize distortions, and a condenser (lens) is used to increase the object illumination.

A microscope will increase the resolving power, i.e. the ability to distinguish close objects, of the unaided eye.

6.3   The astronomical telescope

The object will usually be at infinity, and in normal adjustment the image is at infinity also, so that the telescope length is $f_o + f_e$

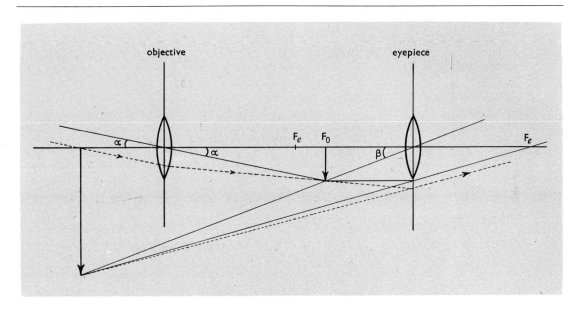

Fig. 57   The astronomical telescope

(If the object is nearer, the length is greater, if the image is nearer, the length is less).

$$\text{Angular magnification} = \frac{\begin{array}{c}\text{Angle subtended at eyepiece by}\\ \text{final image}\end{array}}{\begin{array}{c}\text{Angle subtended at eyepiece by}\\ \text{object}\end{array}}$$

$$\approx \frac{\begin{array}{c}\text{Angle subtended at eyepiece by}\\ \text{final image}\end{array}}{\begin{array}{c}\text{Angle subtended at objective by}\\ \text{object}\end{array}}$$

$$= \frac{\beta}{\alpha}$$

and in normal adjustment

$$= \frac{y}{f_e} \bigg/ \frac{y}{f_o}$$

$$= \frac{f_o}{f_e}$$

y = height of intermediate image

So f_o is longer than f_e.

The resolving power depends upon the diameter of the objective, being greater the larger the diameter. So in high-power telescopes a large diameter objective is required. However, it is difficult to obtain large pieces of flawless glass and make them into lenses. Also the finished lens could only be supported round its edge so distortions would occur due to its own weight. Therefore large telescopes usually have a mirror as an objective as this needs only a flawless surface and can be supported from behind.

As the final image is inverted two other systems are used for terrestrial telescopes.

6.4   The terrestrial telescope

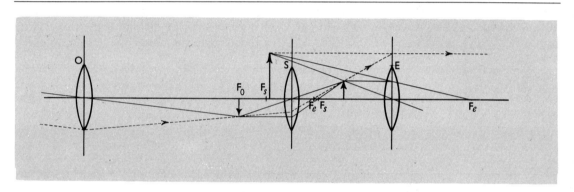

Fig. 58   Terrestrial telescope

The length which is minimum in normal adjustment

$$= f_o + 4f_s + f_e$$

and so the telescope is too long for convenient use.

6.5   The Galilean telescope

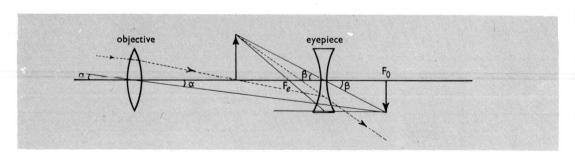

Fig. 59   Galilean telescope

In normal adjustment, the length is $f_o - f_e$, and the angular magnification is still

$$\frac{\beta}{\alpha} = \frac{f_o}{f_e}$$

but for high magnification the field of view is small. The image at F_o is not formed unless the eyepiece is removed.

This system is used for (relatively cheap) opera glasses, but for field glasses etc. the astronomical system, with two erecting prisms, is used to avoid the small field of view of the Galilean system.

MECHANICS

7 MOTION IN A STRAIGHT LINE

7.1 The equations of motion

For motion in a straight line the quantities concerned are:

The initial velocity, u, the velocity at the start of the motion.

The final velocity, v, the velocity at the end of the motion.

The distance travelled, s, during the motion.

The time taken, t, for the motion.

The acceleration, a, the rate of increase of velocity at any point in the motion.

In the simplest case, the acceleration is constant and certain valuable equations arise.

By definition

$$a = \frac{v-u}{t}$$

$$\therefore \quad v = u + at \tag{1}$$

but average velocity

$$= \frac{v+u}{2}$$

(where increase in velocity is uniform)

$$= \frac{s}{t}$$

(from consideration of total distance and time)

$$\therefore \quad s = \frac{u+v}{2}\,t. \tag{2}$$

By simple algebra

$$s = ut + \tfrac{1}{2}at^2 \tag{3}$$

$$s = vt - \tfrac{1}{2}at^2 \quad \text{(This equation is rarely needed)} \tag{4}$$

$$v^2 = u^2 + 2as \tag{5}$$

These five equations will solve all problems of constant acceleration, but the information can often be exhibited very usefully on one or other of two graphs.

7.2 The velocity–time graph

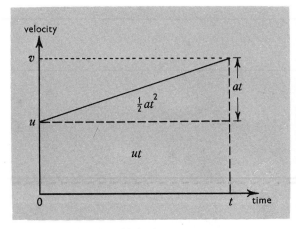

Fig. 60   Velocity–time graph

The area under the graph $= ut + \tfrac{1}{2}at^2 = s$, the distance travelled.

28

The gradient equals the acceleration.

This can be extended to cases of non-uniform acceleration.

7.3   The distance–time graph

Here the gradient equals the velocity.

7.4   The acceleration due to gravity

It can be observed that all objects falling freely near the earth's surface have approximately the same constant acceleration, 'g', the acceleration due to gravity.

$$g \simeq 9 \cdot 8 \text{ m s}^{-2}$$

$$\simeq 980 \text{ cm s}^{-2}$$

7.5   Newton's laws of motion

(*i*)   Every body continues in its state of rest or of uniform motion in a straight line, unless acted upon by an external force.

(*ii*)   Rate of change of momentum is proportional to the applied force and takes place in the direction of the force.

(*iii*)   Action and re-action are equal and opposite.

So from (*ii*)

$$\text{Force, } F \propto \frac{\text{d}}{\text{d}t}(mv)$$

$$\propto m\frac{\text{d}v}{\text{d}t} \text{ if mass is constant}$$

$$\propto ma$$

and units are chosen such that $F = ma$.

Name	Symbol	System	Unit of mass	Unit of acceleration
Newton	N	SI	1 kg	1 m s^{-2}
Dyne	dyn	c.g.s. absolute	1 g	1 cm s^{-2}
Gramme-force	gf	c.g.s. gravitational	1 g	g cm s^{-2}

N.B.   1 gf $= g$ (980) dynes
       1 Newton $= 10^5$ dynes $\simeq 10^2$ g f.

Various pieces of apparatus are now becoming available to verify the relation $F \propto ma$, based on new techniques of reducing friction, and of measuring small time intervals.

7.6   Mass and weight

Mass, once defined too vaguely as the amount of matter in a body, is a constant property measuring a body's resistance to acceleration.

Weight is the force on a body caused by the gravitational attraction of the earth and so varies slightly with position on the surface of the earth and varies considerably as one goes away from the earth.

Weight = Mass × Acceleration due to gravity

or

$$W = mg$$

8 VECTORS, SCALARS AND PROJECTILES

8.1   Vector and scalar quantities

Many physical quantities have a direction associated with them, e.g. force, acceleration, velocity, and they cannot be completely specified unless this direction is known. Such quantities are called vectors. Quantities which do not have an associated direction are called scalars.

8.2   The addition of scalars and vectors

Scalars, being effectively numbers, add arithmetically.

Vectors obey different rules, which are more simply dealt with by drawing.

E.g., find the result of adding a displacement of 4 miles north and 3 miles east.

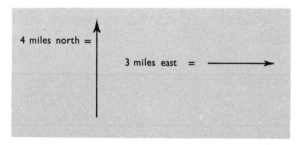

Fig. 61   Vector displacements

They are added by starting with either vector and continuing from where that finishes with the second.

The sum, or resultant, is found by joining starting point to finishing point, and is independent of the order of addition. In this case the resultant is 5 miles, N37·5°E.

All vector quantities obey the same rules, variously labelled as the triangle, parallelogram or polygon of vectors.

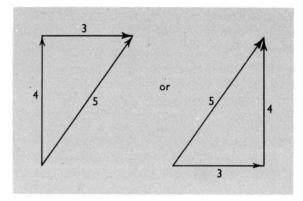

Fig. 62   Vector addition

If the starting and finishing points coincide there is no resultant, which, in the case of forces, is interpreted as equilibrium.

8.3   The resolution of vectors

Very frequently the reverse process is more important and one vector is replaced by two others, usually perpendicular, which are together equal to the original vector.

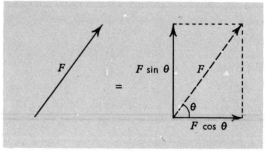

Fig. 63   Resolution of vectors

The two vectors, equal to the original, are referred to as the components of the original vector and the original vector is said to be resolved into its components.

8.4   Projectiles

The study of projectile motion combines the ideas of vectors and the equations of uniform acceleration.

Suppose a particle is projected at an angle α to the horizontal, with velocity v.

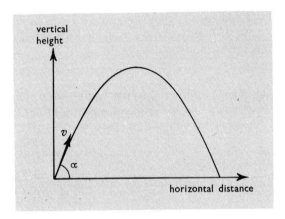

Fig. 64   Projectile motion

Resolve the original velocity into vertical and horizontal components.

Horizontal component $= v \cos \alpha$ and is constant

Vertical component $= v \sin \alpha$ and is subject to a downward acceleration of g.

Consider the vertical motion only and suppose that it takes time t to return to its original vertical height;

use
$$s = ut + \tfrac{1}{2}at^2$$

then
$$0 = v \sin \alpha \,.\, t - \tfrac{1}{2}gt^2$$

$$\therefore \quad t = 0 \text{ or } t = \frac{2v \sin \alpha}{g}$$

In this latter time it travels a horizontal distance
$$v \cos \alpha \cdot \frac{2v \sin \alpha}{g}$$

$$\therefore \quad \text{Horizontal range} = \frac{2v^2 \sin \alpha \,.\, \cos \alpha}{g}$$

$$= \frac{v^2 \sin 2\alpha}{g}$$

Note that the maximum value of $\sin 2\alpha = 1$ and so the maximum range (v^2/g) occurs when $\alpha = 45°$.

9 MOMENTS, COUPLES AND EQUILIBRIUM

9.1   The moment of a force

If a force acts on a body, the body, in general, will tend to move along and to turn.

The tendency of a force to cause turning about any point in the body is measured by the moment of the force about the point. This is defined as the product of the force and the perpendicular distance from the point to the line of action of the force.

i.e. Moment of F about $P = F \times PA$.

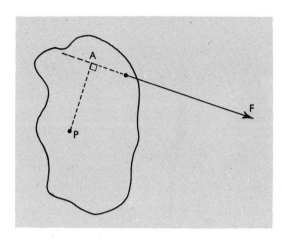

Fig. 65 The moment of a force

9.2   Couples

If two equal and opposite forces are applied to a body there is no tendency for the body to move along, but there is still a tendency for it to turn. Such a system of forces is called a couple.

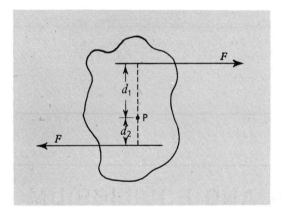

Fig. 66 The torque of a couple

The moment about P is

$$= Fd_1 + Fd_2$$
$$= F(d_1 + d_2)$$

and is thus independent of the position of P. This moment is often called the torque of the couple.

9.3   Equilibrium

If a body remains at rest under the action of a set of forces, it is said to be in equilibrium. The three types of equilibrium may be distinguished by giving the body a small displacement.

(*i*)   If it returns to its original position, it was in stable equilibrium.

(*ii*)   If it remains in its displaced position, it was in neutral equilibrium.

(*iii*)   If it moves away from its original position, it was in unstable equilibrium.

9.4   The conditions for the equilibrium of a rigid body under the action of coplanar forces

(*i*)   The algebraic sum of the resolved components of all forces in any two mutually perpendicular directions must be zero.

(*ii*)   The algebraic sum of the moments of the forces about any point must be zero. (Principle of moments.)

9.5   Centre of gravity

If a body is considered as made up of many small parts, the weights of each part form a set of parallel forces. About one point, the centre of gravity, the sum of the moments of them all is zero. Thus the whole weight may be taken as acting at that point.

Its position may be found by

(*i*)   balancing
(*ii*)   suspension
(*iii*)   calculation

9.6   Centre of mass

This is the point where the total mass of the body appears to act. It will be the same as the centre of gravity for all practical purposes, except in the case of enormous objects where the weights of the component 'particles' are no longer parallel forces.

10   CIRCULAR MOTION

10.1   Uniform motion in a circle

Consider a particle moving uniformly round a circle.

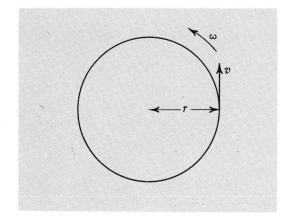

Fig. 67   Uniform motion in a circle

Two velocities may be distinguished.

$$v, \text{the linear velocity} = \frac{\text{Distance round arc}}{\text{Time taken}}$$

$$\omega, \text{the angular velocity} = \frac{\text{Angle turned through}}{\text{Time taken}}$$

v is continually changing direction as the motion is always tangential and this means that there is an acceleration.

ω is specified (fully) as about the axis so is still a vector quantity.

To relate these velocities consider the particle moving in time t a small distance s along the arc, subtending an angle θ radians at the centre.

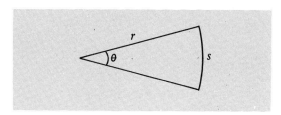

Fig. 68   Angular and linear velocities

From definition
$$v = \frac{s}{t}$$

$$\omega = \frac{\theta}{t}$$

but
$$\theta = \frac{s}{r}$$

$$\therefore \quad \omega = \frac{s}{rt} = \frac{v}{r}$$

or
$$v = r\omega$$

10.2   Acceleration

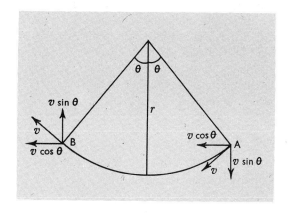

Fig. 69   Angular acceleration

Let the particle move from A to B and introduce a reference radius ending half-way between A and B.

The velocity at A and B is v units tangentially in each case.

Resolve this velocity into components parallel and perpendicular to the reference radius.

For the perpendicular components there is no change in velocity ∴ no acceleration.

For the parallel components the velocity changes by $2v \sin \theta$ and provided θ is taken to be small, this is a radial change.

The change takes place in time

$$\frac{2r\theta}{v}$$

So the acceleration is

$$\frac{2v \sin \theta}{2r\theta/v}$$

but as θ is small $\sin \theta = \theta$

$$\therefore \text{Acceleration} = \frac{v^2}{r} = r\omega^2$$

inwards, along the radius.

10.3   Force

To provide this acceleration a force $mv^2/r = mr\omega^2$ must act on the particle. The inward force acting on the particle is often referred to as the centripetal force. The equal and opposite force (by Newton's third law) acting on the constraint is called the centrifugal force. Note that if the centripetal force vanishes the particle continues to move along a tangent, and this explains, for example, the dangerous breakup of a flywheel under excessive speed.

10.4   Periodic time

This is the time for one complete cycle and is therefore equal to

$$\frac{2\pi r}{v} \quad \text{or} \quad \frac{2\pi}{\omega}$$

10.5   Examples of circular motion

(a)   *The conical pendulum*

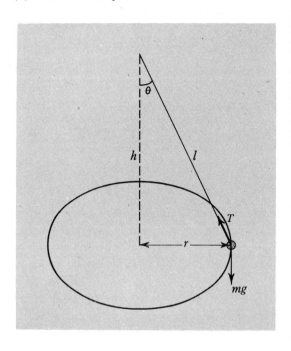

Fig. 70   The conical pendulum

A bob of mass m hangs on a string of length l and describes a horizontal circle of radius r, the string sweeping out a cone of semi-angle θ.

The forces acting are the tension and the weight of the bob.

Resolve T into vertical and horizontal components. The vertical component balances the weight; $T \cos \theta = mg$ The horizontal component supplies the necessary centripetal force to maintain uniform circular motion.

$$T \sin \theta = \frac{mv^2}{r}$$

Divide,

$$\frac{v^2}{rg} = \tan \theta$$

$$v^2 = rg \tan \theta$$

The periodic time,

$$\tau = \frac{2\pi r}{v} = \frac{2\pi r}{\sqrt{rg \tan \theta}}$$

$$= 2\pi \sqrt{\frac{r}{g \tan \theta}}$$

but $\tan \theta = r/h$

$$\therefore \quad \tau = 2\pi \sqrt{\frac{h}{g}} = 2\pi \sqrt{\frac{l \cos \theta}{g}}$$

(b) Equilibrium on a banking

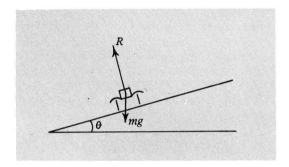

Fig. 71 Equilibrium on a banking

The car of mass m, moves in a horizontal circle of radius r on a banking at angle θ to the horizontal, with a velocity, v. The angle θ is chosen so that no frictional force is needed between the car and the road.

Resolve the reaction, R, into vertical and horizontal components.

The vertical component balances the weight $R \cos \theta = mg$

The horizontal component supplies the necessary centripetal force to maintain uniform circular motion

$$R \sin \theta = \frac{mv^2}{r}$$

Divide

$$\tan \theta = \frac{v^2}{rg} \qquad v = \sqrt{rg \tan \theta}$$

(c) Motion in a vertical circle

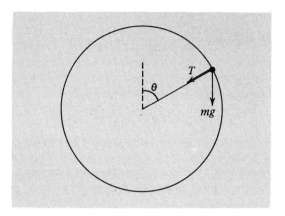

Fig. 72 Motion in a vertical circle

The tension in the string supplies the necessary centripetal force, but the weight of the particle has to be taken into account.

Above the horizontal the weight acts to reduce tension, as it supplies part of the necessary centripetal force.

Below the horizontal it increases the tension.

So the tension varies between $mr\omega^2 \pm mg$ if ω is constant

as $\quad T + mg \cos \theta = mr\omega^2$

If the string breaks; the particle starts off along a tangent, then following the motion of a projectile.

11 WORK, ENERGY AND POWER

11.1 Work

If a force acts on a body and as a result the body moves in the direction of the force, then the force is said to do work on the body.

The amount of work done = force × distance moved in the direction of the force.

If the force and the movement are not along the same line, the force is resolved into components parallel and perpendicular to the movement and the parallel component is used.

$$W = F \cos \theta \times s$$

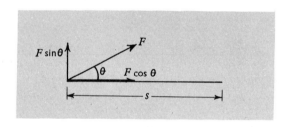

Fig. 73 Work done by a force

11.2 Energy

If a body is capable of doing work then it possesses energy. The amount of energy stored is equal to the total amount of work that can be done. Various forms of energy can be distinguished:

Potential — mechanical energy associated
           with position
Kinetic — mechanical energy associated
           with movement
Heat
Electrical
Chemical

Wave
Mass — includes nuclear energy

Whenever energy is converted from one form to another the total amount before and afterwards is the same, provided every form of energy is considered. This is the fundamental law of the conservation of energy.

11.3 The formulae for potential and kinetic energy

Consider a mass m raised through a height h. The force,

equal to the weight, $= mg$

so the Work done $= mgh$

$$= \text{Potential energy stored}$$

Consider the same mass accelerated by a constant force F (absolute units) from rest to a final velocity v in time t, covering a distance s.

$$\text{Acceleration} = \frac{F}{m} = \frac{v}{t} \qquad \therefore \quad F = \frac{mv}{t}$$

$$\text{Distance } s = \frac{vt}{2} \text{ (equations of uniform acceleration)}$$

$$\therefore \quad \text{Work} = Fs = \frac{mv}{t} \cdot \frac{vt}{2} = \tfrac{1}{2}mv^2$$

$$= \text{Kinetic energy}$$

11.4 Power

Power is the rate of doing work.

11.5   Units

Quantity	SI	c.g.s. absolute	c.g.s. gravitational
Work	joule	erg joule	cm gf
Energy	joule	erg joule kilowatt-hour	cm gf
Power	watt	erg s^{-1} watt	—

1 erg	= 1 dyne cm
1 joule	= 1 newton metre
1 joule	= 10^7 erg
1 kilowatt-hr	= 36×10^5 joule
1 watt	= 1 joule s^{-1}
1 H.P.	= 745·7 watt

11.6   The work done by a couple

Consider two equal and opposite forces acting tangentially to a wheel and turning it through an angle θ (in radians), the forces remaining tangential at all times:

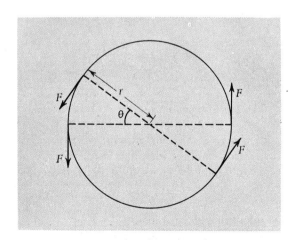

Fig. 74   Work done by a couple

The work done by each force is $Fr\theta$, as $r\theta$ is the length of the arc moved through and at any instant the movement is parallel to the force

$\therefore$ W = Total work = $Fr\theta + Fr\theta$

$\qquad\qquad\qquad = 2Fr\theta$

$\qquad\qquad\qquad$ = Moment of couple $\times \theta$

Thus work done by a couple = moment of couple $\times$ angle of rotation in radians. Whilst this has only been proved in a simple case it can be shown to be generally true.

11.7   The rotational energy of a rigid body

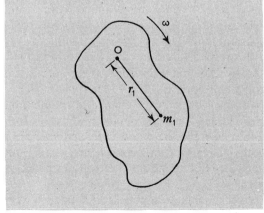

Fig. 75   Rotational energy

Consider a rigid object rotating about an axis through O. This object can be considered to be made up of many particles such as one of mass m_1, distance r_1 from O. Each particle will have the same angular velocity ω, but different speeds $(r_1\omega)$.

Thus the kinetic energy of this particle is $\frac{1}{2}m_1r_1^2\omega^2$. For the whole body, considering it to be made up of many such particles,

The total K.E. $= \frac{1}{2}m_1r_1^2\omega^2 + \frac{1}{2}m_2r_2^2\omega^2$
$$+ \frac{1}{2}m_3r_3^2\omega^2 \ldots .$$

$$= \frac{1}{2}\omega^2(m_1r_1^2 + m_2r_2^2 + m_3r_3^2 \ldots)$$
$$= \frac{1}{2}\omega^2(\Sigma mr^2) = \frac{1}{2}I\omega^2$$

Now the quantity $(\Sigma\, mr^2)$ is very important in cases of rotational motion of a rigid body, and is known as the moment of inertia of the body. It can be calculated for any object about any axis, but this involves integration so for elementary needs it is sufficient to recognise it as a quantity, arising as shown, and dependent on the mass of a body and on the distribution of mass within that body.

12 MOMENTUM

12.1 The conservation of momentum

This quantity has already been met in discussing Newton's Laws of motion, where it was defined as the product of mass and velocity. It is, therefore, a vector quantity, and obeys the important conservation law that: 'In any isolated system the total momentum of all the bodies in a given direction is not changed by any interaction between them'. This is merely a special result from Newton's laws, for consider two objects colliding:

There is a force on each due to the other, and by Newton's third law these forces are equal and opposite. Each experiences, therefore, the same rate of change of momentum (second law) but in opposite senses. Both forces last for the same time and therefore both changes of momentum are equal, but opposite, and therefore the overall change is zero.

Experiments using the new techniques with greatly reduced friction enable elegant verifications to be made.

12.2 Momentum and energy in a collision

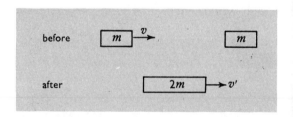

Fig. 76 Momentum and energy in a collision

Consider a mass m, moving with velocity v towards another identical mass initially at rest, and imbedding itself in it. If the combined velocity is v'

Momentum before collision $= mv$
Momentum after collision $= 2mv'$

both in $\rightarrow$ direction

and as momentum is conserved

$$mv = 2mv'$$
$$\therefore \quad v' = \frac{v}{2}$$

K.E. before collision $= \frac{1}{2}mv^2$

K.E. after collision $= \frac{1}{2}(2m)\left(\frac{v}{2}\right)^2$

$$= \frac{1}{4}mv^2$$

So mechanical energy is lost (mainly transformed into heat energy). This is generally true, except in what are called perfectly elastic collisions, when no mechanical energy is lost.

N.B. In this case identical masses interchange velocities.

12.3   Angular momentum

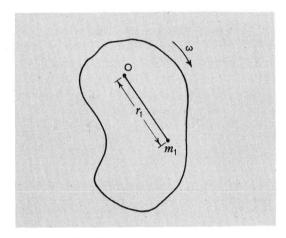

Fig. 77   Angular momentum

Consider the particle m_1 which is part of a rigid body rotating as shown about an axis perpendicular to the paper through O,

$$\text{Linear momentum} = m_1 r_1 \omega$$

$$\text{Moment of momentum} = m_1 r_1{}^2 \omega$$

$$\text{and for the whole body} = \Sigma m_1 r_1{}^2 \omega$$

$$= I\omega$$

called the angular momentum (L).

This is also conserved in an isolated system, e.g. A diver curling up to somersault, or a dancer spinning faster with her arms in to her sides.

12.4   The comparison of linear and angular quantities

$$W = Fs \qquad\qquad W = T\theta$$

$$E = \tfrac{1}{2}mv^2 \qquad\quad E = \tfrac{1}{2}I\omega^2$$

$$P = mv \qquad\qquad L = I\omega$$

$$F = ma \qquad\qquad T = I\frac{d\omega}{dt}$$

13   FRICTION

13.1   The coefficients of friction

Friction is the name given to the force which opposes the motion of one body sliding over another.

Using the apparatus shown, add weights gently to the scale pan. No motion ensues until a certain minimum weight is reached, so the frictional force can take any value up to some maximum value, called the limiting friction.

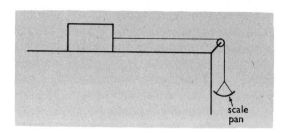

Fig. 78   Limiting friction

It can be shown that the limiting friction

   (*i*)   depends on the material and nature of the surfaces,

   (*ii*)   is proportional to the force pressing the two surfaces together,

   (*iii*)   is independent of the area of contact.

If F_L is the limiting friction, R is the normal reaction, equal to the force pressing the two surfaces together

then $F_L \propto R$

or $F_L = \mu_s . R$

where μ_s is the coefficient of static friction.
   When the block is moving the frictional force becomes F_d, and $F_d = \mu_d R$ where μ_d is the dynamic coefficient of friction.
   It is found that $\mu_s > \mu_d$

13.2   The theory of friction

   The observations above can be explained by considering that the surfaces are molecularly rough so that contact only occurs over a small area ($\simeq 10^{-6}$ of nominal area of contact). Plastic flow will occur under the high pressures created and welding will result. Frictional forces result from the strength of these welds, the extent of which are governed by the loading, not by the area of contact. When slipping occurs the motion is a series of jerks so that the average tension is less than the limiting friction, hence $\mu_d < \mu_s$.

13.3   The measurement of the coefficient of static friction

(*a*)   *Slide and scale pan*

(*b*)   *Inclined plane*

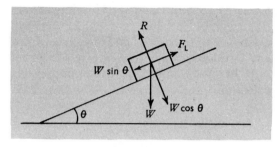

Fig. 79   Block on point of slipping

   Tilt the plane until the block just begins to slip.
   Resolve the weight W into components parallel and perpendicular to the plane.

$$\mu_s = \frac{F_L}{R} = \frac{W \sin \theta}{W \cos \theta} = \tan \theta$$

13.4   Motion down a rough plane

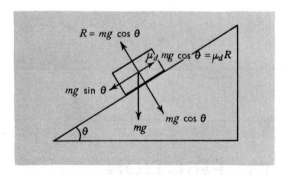

Fig. 80   Block sliding down a plane

   Resolve the weight mg into components parallel and perpendicular to the plane
   Then the normal reaction, $R = mg \cos \theta$, and the limiting friction, $\mu_d R = \mu_d mg \cos \theta$.
   Resultant force down plane $= F$

$$F = mg \sin \theta - \mu_d mg \cos \theta$$
$$= mg(\sin \theta - \mu_d \cos \theta)$$

and acceleration

$$= \frac{F}{m} = g(\sin\theta - \mu_{\mathrm{d}}\cos\theta)$$

13.5   Work done against friction

As friction always opposes motion, if a body moves a distance s, against a frictional force F, the work done is $F \times s$.

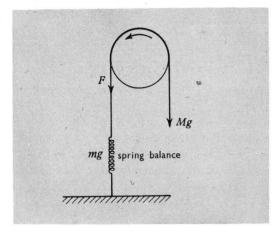

Fig. 81   Measurement of power

This is often used to measure the power of a machine.

A belt is stretched round the pulley connected to the machine. When the pulley is rotated the frictional force plus the tension in the belt (due to the spring balance) balances the tension due to the weight

$$\therefore \quad F + mg = Mg$$

$$F = (M - m)g$$

The belt applies an equal and opposite frictional force to the pulley. If this pulley has radius r, then a point on its circumference moves a distance $2\pi r$ against the force F for each revolution,

thus

$$\text{work done per revolution} = 2\pi rF$$

$$= 2\pi r(M - m)g$$

If the pulley makes n rev sec^{-1} then the power exerted $= 2\pi rng\,(M - m)$

---

14   GRAVITATION

14.1   The theory of gravitation

A knowledge of the theory of gravitation has been developed mainly as a result of astronomers' observations on the movement of the planets.

Great developments were made during the 16th and 17th century. Tycho Brahe made a series of very accurate observations, which Kepler analysed to give three important laws.

1. The planets move in ellipses with the sun at one focus.

2. The line joining the sun and the planet sweeps out equal areas in equal times.

3. The squares of the periods of revolution of the planets are proportional to the cubes of the major axes of their orbits.

These can all be explained by a Law of

Gravitation, first discovered by Newton (1666).

This states that any particle will attract any other particle with a force proportional to the products of the masses and inversely proportional to the square of the distance apart.

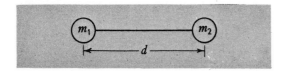

Fig. 82   Law of gravitation

$$F \propto \frac{m_1 m_2}{d^2}$$

$$= \frac{G m_1 m_2}{d^2}$$

where G is the universal gravitational constant

$6 \cdot 67 \times 10^{-11}\,\mathrm{N\,m^2\,kg^{-2}}\ (6 \cdot 66 \times 10^{-8}\,\mathrm{g^{-1}\,cm^3\,s^{-2}})$

14.2   The measurement of G

As the gravitational force between masses of ordinary size is very small, very sensitive methods must be used to measure it.

One well-known method was that of Boys (1895) based on an earlier, larger-scale, method of Cavendish (1798).

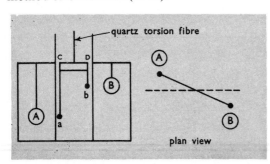

Fig. 83   Boys' apparatus

a,b are identical gold balls 0·2in. dia.
A,B are lead spheres 4·5in. dia.
CD is a highly polished metal bar.

The attractions between the corresponding masses cause a slight movement of the metal bar which can be observed by light reflected from it.

The success of this method depended on the fine, sensitive quartz fibres used. As the apparatus was small, the effect of convection currents, etc. could be more easily controlled.

To calculate G, the deflection is noted and to calibrate the torsion fibre the period of oscillation is observed.

14.3   The relation between G and g

Consider a mass m near the surface of the earth.

If $M =$ mass of earth, $R =$ radius of earth, then as the gravitational attraction provides the weight, and a sphere behaves as though all its mass were concentrated at its centre,

$$\frac{GMm}{R^2} = mg \quad M = \frac{gR^2}{G}$$

If $\rho =$ average density of earth $M = \frac{4}{3}\pi R^3 \rho$

$$\therefore \quad \frac{4}{3}\pi R^3 \rho = \frac{gR^2}{G}$$

$$\rho = \frac{3g}{4\pi RG}$$

and by substitution $\rho \simeq 5500\,\mathrm{kg\,m^{-3}}(5\cdot5\,\mathrm{g\,cm^{-3}})$

14.4   Satellite motion

Consider an object of mass m moving with a linear speed v in a circular path of radius r round a planet mass M. Then as gravitational attraction supplies the necessary centripetal force for uniform circular motion

$$\frac{GMm}{r^2} = \frac{mv^2}{r}$$

$$v^2 = \frac{GM}{r}$$

now periodic time

$$T = \frac{2\pi r}{v}$$

$$\therefore \left(\frac{2\pi r}{T}\right)^2 = \frac{GM}{r}$$

$$4\pi^2 r^3 = GMT^2 \quad \text{(cf. Kepler's third law)}$$

14.5 Escape velocity

If an object is projected at a high velocity it may have sufficient energy to escape from a planet. The minimum velocity necessary for this is called the escape velocity.

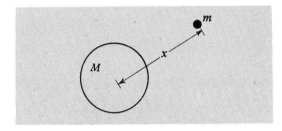

Fig. 84   Escape velocity

Consider the mass m at a distance x from the centre of the planet of mass M.

Force acting $= F = \dfrac{GMm}{x^2}$

Let object move a distance δx

Work done, $\delta W = \dfrac{GMm\delta x}{x^2}$

In escaping, work done $= \displaystyle\int dW = \int_R^\infty \frac{GMmdx}{x^2}$

R = radius of planet $\quad = \left[\dfrac{-GMm}{x}\right]_R^\infty$

$$= \frac{GMm}{R}$$

Now if this comes entirely from the original kinetic energy

$$\tfrac{1}{2}mv^2 = \frac{GMm}{R}$$

$$v^2 = \frac{2GM}{R}$$

and as $M = \dfrac{gR^2}{G}$ (See §14.3)

$$v^2 = 2gR$$

$$\therefore \quad v = \sqrt{2gR}$$

and by substitution $\simeq 11 \text{ km s}^{-1}$

15   SIMPLE HARMONIC MOTION

15.1 The circle method

Consider a particle moving with constant angular velocity ω, round a circle. Project the motion on to a diameter:

Then as shown the acceleration of the pro-jection is equal to the component of the acceleration of the particle parallel to the diameter

$$= \omega^2 r \sin \theta$$

$$= \omega^2 x$$

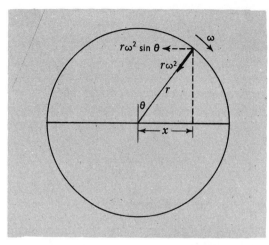

Fig. 85   Simple harmonic motion

Now the acceleration is always directed to the centre of the circle, and so the projection performs a new and important type of motion, called simple harmonic motion (s.h.m.), defined as:

Motion such that the acceleration is proportional to the distance from a fixed point and directed towards that point.

Whilst this is a somewhat artificial example it enables the properties of the motion to be discussed before going on to other examples.

The amplitude is the maximum displacement ($= r$)

The periodic time is the time for one complete cycle ($= 2\pi/\omega$)

The frequency is the number of cycles per second ($= \omega/2\pi$)

The velocity of the projection equals the component of the velocity of the particle parallel to the diameter

$$= r\omega \cos \theta$$

but

$$\cos \theta = \frac{\sqrt{r^2 - x^2}}{r}$$

$$\therefore \quad \text{vel}^2 = \omega^2(r^2 - x^2)$$

If a particle moving with the projection has mass m, the force acting on it $= m\omega^2 x$, and

at unit displacement the force acting $= m\omega^2$

$$\therefore \quad \omega^2 = \frac{\text{Force at unit displacement}}{\text{Mass moving}}$$

so

$$\text{Periodic time} = \frac{2\pi}{\omega}$$

$$= 2\pi \sqrt{\frac{\text{Mass moving}}{\text{Force at unit displacement}}}$$

This is a general formula which will be used in the following examples.

15.2 A mass on the end of an elastic cord or spring

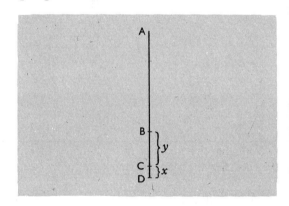

Fig. 86   Motion of a mass on a spring

Consider a spring of unloaded length AB which stretches to C when a steady load (mg) is added. If the system is given a further displacement to D and then released it will oscillate about C.

It is assumed:

(*i*)   That the mass of the spring can be ignored,

(*ii*)   That the motion is in a vertical line,

(*iii*)   That the amplitude of the motion is less than the static extension, so that the spring never becomes slack,

(*iv*)   That all extensions obey Hooke's law,

(*v*)   That air resistance and any other source of energy loss can be ignored.

Then if $BC = y$ and $CD = x$

In equilibrium

$$y \propto mg \quad \text{(Hooke's law)}$$

$$\therefore \quad mg = ky \quad k = \text{spring constant}$$

When stretched the further distance x, the force due to the spring $= k(y+x)$

i.e.

$$\text{Restoring force} = k(y+x) - mg$$

$$= kx \quad \text{upwards}$$

which is proportional to x and directed to the equilibrium position

$\therefore$　s.h.m. and the force at unit displacement

$$= k = \frac{mg}{y}$$

so periodic time

$$= 2\pi \sqrt{\frac{\text{mass moving}}{\text{force at unit displacement}}}$$

$$= 2\pi \sqrt{\frac{m}{mg/y}} = 2\pi \sqrt{\frac{y}{g}}$$

For a standard experiment, transpose the formula

$$t = 2\pi \sqrt{\frac{y}{g}} \quad \text{to} \quad g = \frac{4\pi^2 y}{t^2} = 4\pi^2 \left(\frac{y}{m}\right)\left(\frac{m}{t^2}\right)$$

And plot two graphs:

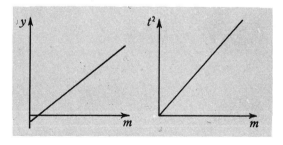

Fig. 87　Graphs for the behaviour of a mass on a spring

The gradients of these graphs are used to calculate a value for g.

15.3　The simple pendulum

This is formed by attaching a heavy mass to a light cord and allowing it to oscillate in a vertical plane.

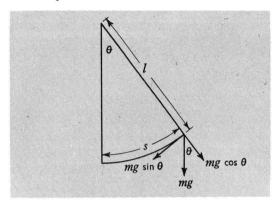

Fig. 88　Simple pendulum

It is assumed:

(*i*)　That the weight of the cord may be neglected.

(*ii*)　That the motion is in a vertical plane.

(*iii*)　That air resistance and any other source of energy loss can be ignored.

(*iv*)　That the upper support does not move.

(*v*)　That no energy is required to bend the cord.

(*vi*)　That θ is small (< 5 or $6°$).

Then the forces acting on the bob are the tension in the cord and the weight. Resolve the weight into components parallel and perpendicular to the cord.

The parallel component $mg \cos \theta$ balances the tension.

The perpendicular component, $mg \sin \theta$ supplies a restoring force

$$mg \sin \theta = mg\,\theta \quad \text{as } \theta \text{ is small}$$

$$= \frac{mgs}{l}$$

So the restoring force is proportional to the displacement and directed to the centre of the motion

$$\therefore \quad \text{s.h.m.}$$

The restoring force at unit displacement = mg/l and

$$\text{Periodic time} = 2\pi \sqrt{\frac{\text{mass moving}}{\text{force at unit displacement}}}$$

$$= 2\pi \sqrt{\frac{m}{mg/l}} = 2\pi \sqrt{\frac{l}{g}}$$

15.4 A pendulum with an inaccessible suspension point

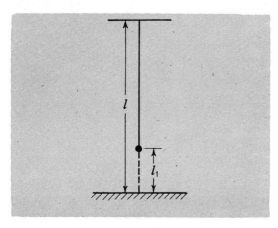

Fig. 89 Pendulum with inaccessible suspension point

If the upper support of the pendulum is inaccessible the distance from bob to floor is measured, and then

$$t = 2\pi \sqrt{\frac{l - l_1}{g}}$$

$$t^2 = \frac{4\pi^2 l}{g} - \frac{4\pi^2 l_1}{g}$$

Plot t^2 against l_1, and g and l can be determined from the graph:

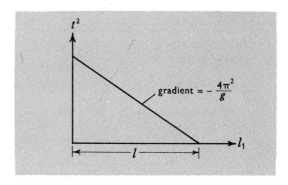

Fig. 90 Graph for pendulum with an inaccessible suspension point

15.5 The effect of changes in gravity on an oscillating spring and a simple pendulum

(a) *Spring*

$$t = 2\pi \sqrt{\frac{y}{g}} \quad \text{but} \quad y \propto g \quad \therefore \quad t \text{ constant if } g$$
changes

(b) *Pendulum*

$$t = 2\pi \sqrt{\frac{l}{g}} \quad l \text{ is constant} \quad \therefore \quad t \text{ varies if } g$$
changes

16   HYDROSTATICS

This studies effects in fluids at rest.

16.1   Fluid pressure

In a fluid at rest the pressure is normal to the surface. The pressure is the force per unit area at a point in the fluid.

16.2   The pressure at a point is the same in all directions

Consider a wedge of infinitesimal size in a fluid, so the weight of the fluid in the wedge may be neglected. Let the pressures on the three faces be as shown.

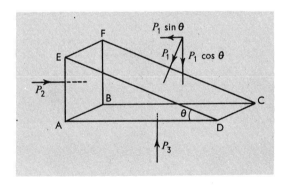

Fig. 91   Pressure at a point in a fluid

If area of face $CDEF = A$

$\therefore$ area of face $ABCD = A \cos \theta$

area of face $ABFE = A \sin \theta$

Now total horizontal and vertical forces must both be zero or else there would be movement. Resolve $P_1 A$ into horizontal and vertical components.

Equating horizontal forces

$$P_1 A \sin \theta = P_2 . A \sin \theta$$

$$\therefore \quad P_1 = P_2$$

Equating vertical forces

$$P_1 . A \cos \theta = P_3 . A \cos \theta$$

$$P_1 = P_3$$

so $P_1 = P_2 = P_3$ $\therefore$ Pressure at a point is the same in all directions.

(a) *The pressure at all points at the same horizontal level in a fluid is the same*, or else there would be fluid flow under the horizontal pressure gradient.

(b) *A fluid transmits an externally applied pressure increment equally in all directions.* This follows as a result of pressure at a point being the same in all directions and Newton's third law. Any excess must be transmitted to all parts of the fluid.

16.3   The hydraulic (Bramah) press

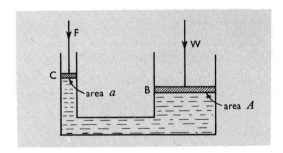

Fig. 92   The hydraulic press

This is an application of the last result.

$$\text{Pressure at C} = \frac{F}{a}$$

and this equals

$$\text{Pressure at B} = \frac{W}{A}$$

so

$$W = \frac{F.A}{a},$$

hence W can be very much larger than F.

16.4   The pressure at a given depth in a liquid

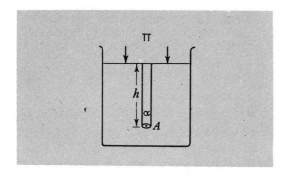

Fig. 93   The pressure in a liquid

The pressure at some point A, at a depth h, is due to the weight of the liquid above it, plus the pressure on the free surface.

Imagine a column of liquid of cross-sectional area α.

$$\text{Weight} = \alpha h \rho g$$

$$\text{so excess pressure} = \text{weight/area}$$

$$= g\rho h$$

$$\text{Total pressure at A} = \Pi + g\rho h$$

This is also the expression for the pressure at the bottom of a liquid column of height h.

16.5   The manometer

Pressure at A = Pressure at B as on same horizontal level.

Pressure at B = Pressure at C + Pressure due to column of liquid, height h.

If the density of the liquid is ρ.

Pressure at A   = Atmospheric pressure
                     $+ h$ (mm of liquid)
       = Atmospheric pressure
                     $+ g\rho h$

which can be expressed in N m^{-2} (or dyn cm^{-2})

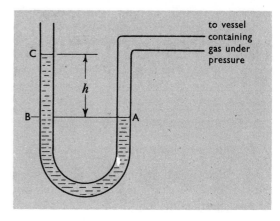

Fig. 94   The manometer

providing atmospheric pressure is measured in the corresponding units.

16.6   The barometer

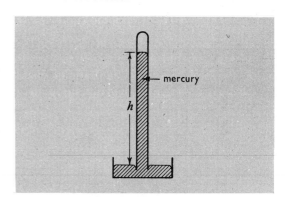

Fig. 95   The barometer

Atmospheric pressure = h mm of mercury

Standard atmosphere = 760 mm of mercury

$$= 1 \cdot 013 \times 10^5 \text{ N m}^{-2}$$

$$= 1 \cdot 013 \times 10^6 \text{ dyn cm}^{-2}$$

$$= 1013 \text{ millibars}$$

(1 bar = 10^6 dyn cm^{-2} = 10^5 N m^{-2})

There are two fundamental types of barometer, mercury and aneroid. The aneroid depends on the flexing of partially evacuated metal bellows under pressure changes. It must be calibrated with reference to a mercury barometer. It can be equally accurate and is independent of g.

16.7   The Fortin barometer

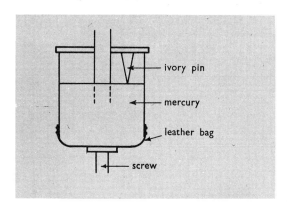

Fig. 96   The lower part of a Fortin barometer

The mercury level can be adjusted so as to be always in contact with the tip of an ivory pointer which is the zero of the scale.

16.8   Hare's apparatus

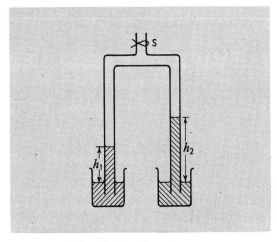

Fig. 97   Hare's apparatus

If the pressure is reduced at S by sucking, the liquids will rise up to different heights so that the difference in pressure between outside and inside equals

$$g\rho_1 h_1 = g\rho_2 h_2$$

Thus
$$\rho_1 h_1 = \rho_2 h_2$$
$$h_1 = \frac{\rho_2}{\rho_1} h_2$$

h_1 is plotted against h_2, and the ratio of the densities found from the gradient.

Wide, wetted tubes are used to avoid surface tension effects.

16.9   Archimedes' principle

When a solid body is wholly or partly immersed in a fluid, then it experiences an upthrust equal to the weight of the displaced fluid. This upthrust acts vertically through the centre of gravity of the displaced fluid.

The connection between Archimedes' principle and the pressure in a liquid can be shown for a simple case as follows.

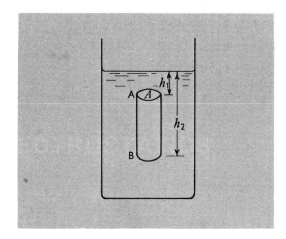

Fig. 98   Archimedes' principle

Consider a cylinder of cross-sectional area A immersed in a liquid of density ρ.

Pressure at A $= g\rho h_1$,
Force on face at A $= g\rho h_1 A$ (downwards)
Pressure at B $= g\rho h_2$,
Force on face at B $= g\rho h_2 A$ (upwards)
So difference in force
$\qquad\qquad$ = upthrust
$\qquad\qquad$ = $g\rho A (h_2 - h_1)$ (upwards)
$\qquad\qquad$ = weight of fluid displaced

16.10 The experimental verification of Archimedes' principle

(*i*) Bucket and Cylinder Method. Fairly accurate, but only gives one reading.

(*ii*) Eureka Can Method. Several readings, but only poor accuracy.

16.11 The principle of flotation

A floating body displaces its own weight of fluid.

16.12 The constant weight hydrometer

Let the volume to the base of the stem $= V$

     the cross-sectional area of the stem $= A$,

     and the weight of the hydrometer $= W$.

Let it float so that a length x of stem is immersed in a liquid of density ρ

Then volume of liquid displaced $= V + Ax$

Weight of liquid displaced $= (V + Ax)\rho g$

by Principle of Flotation $= W$

$$\therefore \quad \rho = \frac{W}{(V + Ax)g}$$

Thus the scale is non-linear, and the smaller A the greater the change in x must be for the same change in ρ, i.e. the instrument is more sensitive.

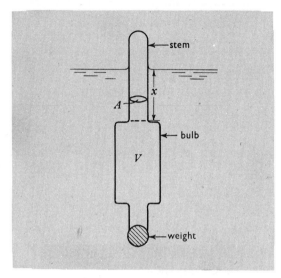

Fig. 99   The hydrometer

17   MOTION IN FLUIDS

17.1 Viscosity

When a liquid moves through a tube or an object moves through a liquid, there is a resistance to motion caused by the liquid and said to be due to its viscosity (cf. glycerine and water).

The flow of liquid through a tube, and the motion of a ball bearing through a viscous liquid will be considered qualitatively.

17.2 The flow of liquid through a tube

(*a*) *Experiment 1*

For slow rates of flow potassium permanganate injected into the water shows

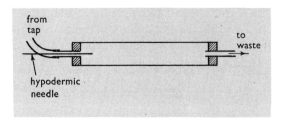

Fig. 100 Streamline and turbulent flow

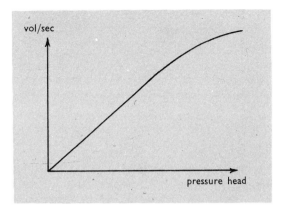

Fig. 101 Flow of liquid through a tube

straight linear coloration, and the liquid is said to be moving with streamline flow.

For higher rates of flow the pattern breaks up irregularly, and the liquid is said to be moving with turbulent flow.

17.3 The motion of a sphere in a viscous liquid

(c) *Experiment 3*

If the velocity of a small ball bearing is measured as it falls through a viscous liquid, it will be found that the velocity soon reaches a steady value, the terminal velocity, when the viscous drag balances the (accelerating) weight.

(b) *Experiment 2*

If the volume of liquid flowing per second through a capillary tube is plotted against the pressure drop across the tube, the resulting graph shows that there is a linear relationship (whilst the flow is streamline) followed by a non-linear relationship (whilst the flow is turbulent).

---

18   ELASTICITY

This is the study of the behaviour of bodies when acted upon by forces which tend to cause deformation.

18.1 The behaviour of a stretched wire (general case)

L — Limit of proportionality. Up to this point Hooke's law is obeyed, i.e. extension is proportional to load.

E — Elastic limit. Up to this point the wire will return to its original length if unloaded. If taken beyond this point and unloaded it will have a permanent increase in length, called a permanent set.

Y — Yield point. Beyond this point deformation is uniform along the wire, but time dependent.

N — Necking point. The wire begins to thin, or form a neck, at one point.

B — Breaking point. Where the wire snaps, normally within the neck.

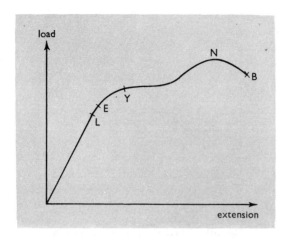

Fig. 102   Typical load–extension graph for a metal

This curve will vary from metal to metal and is more typical of a ductile material. A brittle material will snap without necking.

18.2   Stress and strain

As area is important these quantities are introduced to replace load and elongation.

$$\text{Stress} = \frac{\text{Force}}{\text{Cross-sectional area}}$$

$$\text{Strain} = \frac{\text{Change in length}}{\text{Original length}}$$

18.3   Young's modulus

Up to the limit of proportionality, strain is proportional to stress,

$$\therefore \quad \frac{\text{Stress}}{\text{Strain}} = \text{constant, Young's modulus of elasticity}$$

$\therefore \quad$ Young's modulus of elasticity

$$= \frac{\text{Force} \times \text{Original length}}{\text{Area} \times \text{Extension}}$$

18.4   The determination of Young's modulus

Two identical wires of the same material are fastened to a rigid support. One wire carries a scale and fixed load, the other a vernier and scale pan. This method of support allows for yielding of the upper support as a whole and for temperature effects.

The zero reading is noted, weights added to the scale pan, the new vernier reading noted. Then the weights are removed and the zero checked to see if anything has slipped or the elastic limit has been passed. This is repeated in suitable steps appropriate to the wire.

The diameter of the wire is measured with a micrometer screw gauge in mutually perpendicular directions at three points on the wire. The original length is also found separately with a metre rule.

The above formula can then be used, finding the average value of load/extension from a graph.

18.5   The work done in stretching a wire

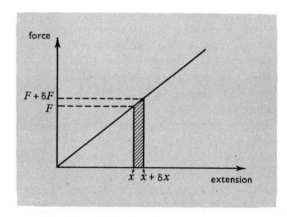

Fig. 103   Work done in stretching a wire

$$\text{Work done} = \text{Force} \times \text{Distance}$$

so that if a force change F to $F + \delta F$ causes an extension change x to $x + \delta x$.

$$\text{Work done} = \text{Force} \times \text{Distance moved}$$
$$= F\,\delta x$$
$$= \text{Area shaded}$$

so in whole extension

$$\text{Work done} = \text{Area under graph}$$

and if the graph is a straight line

$$\text{Work done} = \tfrac{1}{2}\,\text{Maximum force} \times \text{Total extension}$$

19 SURFACE TENSION

19.1 Elementary phenomena

Many elementary phenomena point to the existence of surface tension. e.g.

    1. The hairs of a wet brush are pulled together when the brush is removed from the liquid.

    2. The spherical shape of small mercury drops.

    3. A needle can be floated on water.

19.2 The molecular theory

A molecule well away from the surface of the liquid receives attractions in all directions from neighbouring molecules. Over an interval of time these attractions cancel out, so there is zero average force on the molecule.

For those near the surface however this is not the case and hence they tend to be pulled towards the bulk of the liquid, tending to make the surface as small as possible.

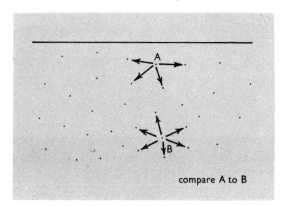

Fig. 104   Theory of surface tension

19.3 The definition of the coefficient of surface tension

XY is a wire free to slide on a wire frame-

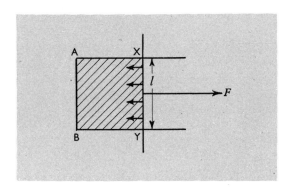

Fig. 105   Coefficient of surface tension

work, and in the space ABYX a liquid surface is included. A force F has to be applied to balance the surface tension force of the liquid surface. If γ is the coefficient of surface tension, then

$\gamma l = F$, or $2\gamma l = F$ if the surface is a film with effectively two surfaces.

Thus the coefficient of surface tension of a liquid is the force acting (within the surface) per unit length at right angles to the boundary tending to make the surface area a minimum.

    The units are N m^{-1} or dyn cm^{-1}.

19.4 Drops and bubbles

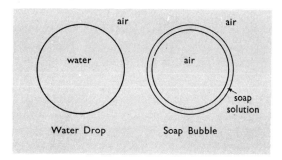

Fig. 106   Drop and bubble

E

A drop has one surface, whereas a bubble has two surfaces.

19.5   The excess pressure inside a drop

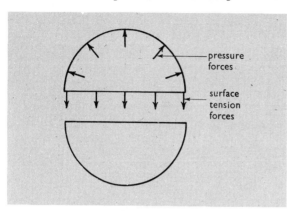

Fig. 107   Excess pressure in a drop

Imagine a drop of radius r, sliced through the middle. If the internal excess pressure is P, the force tending to remove the upper hemisphere is the vertical component of

$$P \times \text{area} = P\pi r^2$$

The force retaining the hemispheres is $\gamma \times 2\pi r$, where γ is the coefficient of surface tension, and as the drop is in equilibrium

$$P\pi r^2 = 2\pi r\gamma$$

$$\therefore \quad P = \frac{2\gamma}{r}$$

19.6   The excess pressure inside a bubble

Imagine a bubble as two concentric drops,

$$\text{Excess pressure due to outer drop} = \frac{2\gamma}{r_2}$$

$$\text{Excess pressure due to inner drop} = \frac{2\gamma}{r_1}$$

$$\therefore \quad P' = \frac{2\gamma}{r_2} + \frac{2\gamma}{r_1} = \text{total excess pressure}$$

but for a bubble $r_1 \simeq r_2$

$$\therefore \quad P' = \frac{4\gamma}{r}$$

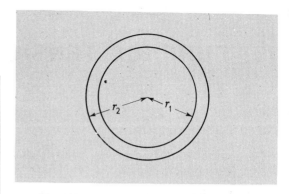

Fig. 108   Excess pressure inside a bubble

Thus the excess pressure is inversely proportional to the radius. This can be demonstrated as follows:

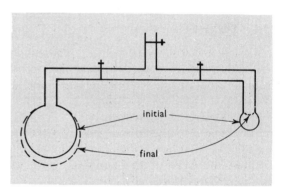

Fig. 109   Growth of bubbles

Two unequal bubbles are blown as shown. When put into communication it is the small one that contracts.

19.7   The angle of contact

The angle between the tangent to the surface of a liquid at the point of contact with a wall, and the wall itself, is called the angle of contact.

The value of the angle depends both upon the solid and the liquid, e.g.

water–glass   0°   i.e. water wets glass
water–chromium   160°
mercury–glass   135°

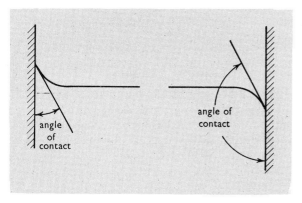

Fig. 110   Angle of contact

19.8   The measurement of the angle of contact

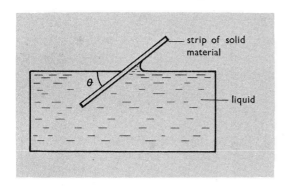

Fig. 111   Measurement of angle of contact

The liquid is placed in a vessel and the strip of the relevant solid material is tilted through an angle. When there is no distortion of the liquid surface right to the point of contact θ is the angle of contact required.

19.9   The measurement of the coefficient of surface tension

The prime requirement in all these methods is to have the apparatus scrupulously clean, as contamination, particularly with grease, seriously affects the surface tension.

(a)   Searle's torsion balance

When the glass plate is just about to break

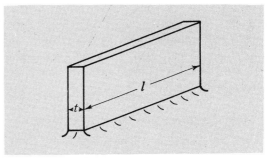

Fig. 112   Microscope slide breaking from a liquid surface

away from the liquid surface the upward force must equal

$$2\gamma\,(l+t) + mg \text{ where } mg$$
$$= \text{the weight of the slide}$$

Hence by measuring this force γ can be found. This can be done using a chemical balance, but more accurately with Searle's torsion balance.

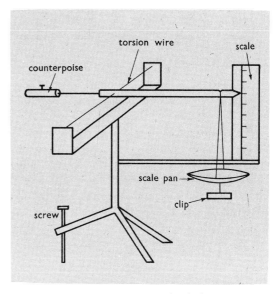

Fig. 113   Searle's torsion balance

The microscope slide is hung from a clip. By adjusting the counterpoise, and turning the screw, the reading on the scale when the slide leaves the liquid uniformly can be found. This same reading can be obtained when the liquid is removed, but with the dry microscope slide still in place, by adding weights to the scale pan. The torsion comes from the wire.

If the required mass is m.

$mg = 2\gamma(l+t)$, the weight of the
$\qquad\qquad\qquad$ slide being eliminated.

A similar type of experiment can be performed using a wire frame instead of a glass slide.

(b) Capillary tube method

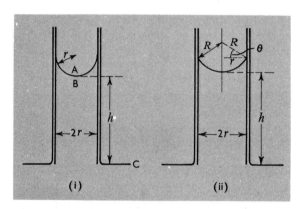

Fig. 114   Capillary rise

In (i) there is zero angle of contact.
The pressure difference between A and B is the same as that across a drop of radius r, i.e $2\gamma/r$, but pressure at A = pressure at C (atmospheric), and pressure difference between B and C is that due to a column of liquid h high $= g\rho h$.

$$\therefore \quad \frac{2\gamma}{r} = g\rho h$$

$$\gamma = \frac{g\rho hr}{2}$$

In (ii) the angle of contact is θ, and $r/R = \cos\theta$

$$\therefore \quad R = \frac{r}{\cos\theta}$$

$$\gamma = \frac{g\rho hr}{2\cos\theta}$$

In practice h is measured by a travelling microscope and r is measured by breaking the tube at the meniscus level and using the travelling microscope. If the tube is clean the liquid will move freely as the tube is moved. If tube is too short a curved surface is formed at the top, of different radius to the tube.

(c) Jaeger's method

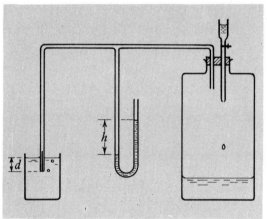

Fig. 115   Jaeger's method

The maximum pressure shown by the manometer as the drops break away singly and slowly is observed.

The drop pressure will be greatest when the radius is smallest. This is when it equals r, the radius of the capillary tube.

Then if ρ_1 is the density of the liquid under test, ρ_2 that of the liquid in the manometer,

$\therefore$ the pressure difference due to surface tension

$$= g(\rho_2 h - \rho_1 d)$$

and

$$= \frac{2\gamma}{r}$$

so

$$\gamma = \tfrac{1}{2}rg(\rho_2 h - \rho_1 d)$$

Full analysis shows this to be over-simplified, and so the method is not usually employed for absolute determinations, but for study of the effect of temperature on surface tension. The apparatus is calibrated at one temperature where the surface tension is known.

19.10 The force between the plates with liquid in between them

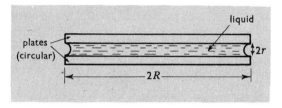

Fig. 116   Force between plates

Here the curved surface is not spherical but of different radius in perpendicular directions and the excess pressure across the surface is equal to

$$\frac{\gamma}{r} - \frac{\gamma}{R}$$

Hence pressure inside is less than atmospheric by

$$\frac{\gamma}{r} - \frac{\gamma}{R}$$

and if the plates, or liquid area is circular of radius R,

$$\text{Force} = \pi R^2 \left(\frac{\gamma}{r} - \frac{\gamma}{R} \right)$$

and as r is usually small the force is usually large.

20   DIMENSIONS

20.1   The dimensions of a quantity

All mechanical quantities can be simply related to the three basic dimensions of mass, length and time. If these three are represented as [M], [L] and [T], then area can be represented as $[L^2]$, acceleration as $[LT^{-2}]$, force as $[MLT^{-2}]$, etc.

The dimensions of a quantity are useful in two ways.

20.2   Checking an equation

All the terms in an equation must have the same dimensions.

e.g. consider the equation

$$s = ut + \tfrac{1}{2}gt^2$$

writing down the dimensions

$$[L] = [LT^{-1}][T] + [LT^{-2}][T]^2$$

So all the terms are dimensionally the same, [L]. Thus the equation is possible, but the method cannot test whether it is actually correct as the numbers have no dimensions and therefore cannot be checked. However an incorrect equation should be revealed.

20.3 Predicting the form of an equation

If it is thought that four variables are related, the method of dimensions can predict the form of the relationship, but not the numbers of dimensionless constants in the equation. e.g. For a simple pendulum the periodic time (t) might be thought to depend on the mass of the bob (m), the length of the suspension (l) and the acceleration due to gravity (g). This can be generally expressed as

$$t = km^{\alpha}l^{\beta}g^{\gamma}$$

where k is a dimensionless constant.

The dimensions can then be written

$$[T] = [M]^{\alpha}[L]^{\beta}[LT^{-2}]^{\gamma}$$

and as the dimensions of both sides of the equation must be the same, the powers can be equated.

Thus, equating powers of

$$[M] \quad 0 = \alpha$$
$$[L] \quad 0 = \beta + \gamma$$
$$[T] \quad 1 = -2\gamma$$
$$\therefore \quad \gamma = -\tfrac{1}{2}, \quad \beta = +\tfrac{1}{2}, \quad \alpha = 0$$

so the relationship must be $t = k\sqrt{l/g}$ but the constant k cannot be found by this technique.

HEAT

21 TEMPERATURE

21.1 Temperature and heat

Temperature is the scientific concept which corresponds to the looser terms 'hot' and 'cold'. Heat energy is that which causes a rise in temperature or a change of state.

A is at a higher temperature than B, if, when A and B are placed in thermal contact, heat flows from A to B.

21.2 The measurement of temperature

In order to give a further meaning to temperature, the following are needed:

(*i*) A numerical scale to which it can be referred.

(*ii*) Two reference temperatures (fixed points) on which to base the temperature scale.

(*iii*) Some property which changes with temperature,

Let temperature be based on the value which this varying property X takes, assuming temperature to be a quantity which varies linearly with X. Let the fixed temperatures be 0 and 100 (so working on the Celsius or Centigrade scale—Celsius being merely a new name to achieve international uniformity).

Then let the value of X at 0°C be X_0
and let the value of X at 100°C be X_{100}
and let the value of X at t°C be X_t

where t is some measured temperature whose value is to be found on the scale.

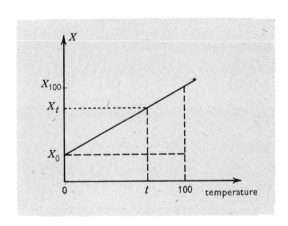

Fig. 117   A scale of temperature

Then by similar triangles

$$\frac{X_{100} - X_0}{100} = \frac{X_t - X_0}{t}$$

$$t = \frac{100 \cdot (X_t - X_0)}{(X_{100} - X_0)}$$

59

21.3   The fixed points

(a)   Lower

A mixture of finely crushed ice and air saturated water at standard atmospheric pressure $1 \cdot 013 \times 10^5$ N m^{-2}.

(b)   Upper

Steam above boiling distilled water at standard atmospheric pressure.

In both cases corrections can be applied if the determination is made at non-standard pressures, using various formulae. At the upper point a one Celsius degree change corresponds to about 27 mm of mercury change in pressure.

21.4   Liquid-in-glass thermometers

The choice of both liquid and glass is most important. Various special glasses distinguished by coloured stripes can be used. The property X is the length of the liquid column from some convenient reference point. However, such thermometers are not used where the highest accuracy is required.

21.5   The choice of liquid

(a)   Mercury

Advantages

(i)   Opaque, so easily seen in a fine tube.

(ii)   Does not wet tube, so none left behind as temperature falls.

(iii)   Long range (m.p.-39°C, b.p. 357°C)

Disadvantages

(i)   Sometimes jerky in movement as it does not wet glass

(ii)   Smaller expansion co-efficient than alcohol

(b)   Alcohol

Advantages

(i)   No jerky movement.

(ii)   Expansion six times that of mercury.

(iii)   Melting point −117°C.

Disadvantages

(i)   Boils at 78°C.

(ii)   Wets tube.

(iii)   Distils to colder parts of tube.

21.6   Special types of liquid-in-glass thermometers

(i)   Clinical thermometer which has a constriction to prevent the return of the mercury to the bulb.

(ii)   Maximum and minimum thermometers, e.g. Rutherford's pattern with an index to record the temperatures, or Six's combined pattern.

(iii)   Earth thermometer which is heavily protected, and sluggish in action, so there is little change while being removed and read.

(iv)   Deep sea thermometer, which is reversed to break the mercury column at the required depth. The volume of the separated mercury can afterwards be found and hence the temperature.

(v)   Beckmann thermometer, where a reservoir is used to give a variable range.

21.7   The range of liquid-in-glass thermometers

Mercury   from   −39°C
          to   300°C
   or   to   500°C with nitrogen
                    above the liquid.
Alcohol   from   −117°C
          to   60°C
Liquid Pentane down to −160°C.

21.8   Other types of thermometer

(i)   Bimetallic thermometers. These are used chiefly in recording instruments

(ii)   Mercury-in-steel thermometers. The pressure of mercury works a Bourdon type gauge. It can have a long connecting tube (to 40 m) to connect thermometer and recorder.

(*iii*) Vapour pressure thermometers. These are very sensitive over a small range.

To be considered in more detail later.

(*iv*) Constant volume air thermometer (X = pressure). See §22.7.

(*v*) Thermo-electric thermometer (X = e.m.f.). See §41.1.

(*vi*) Resistance thermometer (X = resistance). See §40.10(h).

(*vii*) Pyrometers. These instruments depend on radiation laws. See §28.8.

21.9   The international temperature scale (1927 revised 1948)

This is essentially a practical reference scale.

It depends on (*i*) a number of fixed and reproducible equilibrium temperatures to which precise numerical values are assigned, (*ii*) definite formulae and means for interpolation.

(*i*)   Fixed points are:

b.p. oxygen	$-182{\cdot}970°C$	b.p. sulphur	$444{\cdot}600°C$
m.p. ice	$0°C$	f.p. silver	$960{\cdot}8°C$
b.p. water	$100°C$	f.p. gold	$1063{\cdot}0°C$

(*ii*)   Instruments are:

From -182 to $0°C$	Platinum resistance thermometer ⎫ Different
From 0 to $630{\cdot}5°C$	Platinum resistance thermometer ⎬ formulae
From $630{\cdot}5$ to $1063°C$	Platinum: platinum–rhodium thermocouple
From $1063°C$	Pyrometers.

21.10   The variation of the number assigned to a fixed temperature

If the same temperature is measured with different instruments, and the results calculated on the assumption that each instrument has an exactly linear relationship between its property X and temperature, the numerical values for the same temperature will be slightly different.

22   THE GAS LAWS

22.1   Boyle's law

The volume of a fixed mass of gas at constant temperature is inversely proportional to the pressure

i.e.

$$V \propto \frac{1}{P}$$

or

$$PV = \text{constant}$$

22.2 Experimental verification

The simplest apparatus is the J-tube containing mercury, but this is limited to pressures greater than atmospheric (i.e. approximately 1 to 2 atmospheres). To give a wider range of investigation, some modification is necessary, but the principle is still the same.

The gas is dried and enclosed in a graduated tube over mercury. The mercury also fills a rubber tube leading to the movable reservoir.

By altering the position of this reservoir the pressure on the gas is varied, being equal to $\pi + h$ where π is atmospheric pressure (in m of mercury). If the reservoir level is lower than that in the tube the pressure becomes $\pi - h$, therefore allowing an investigation over the range $\frac{1}{2}$ to 2 atmospheres.

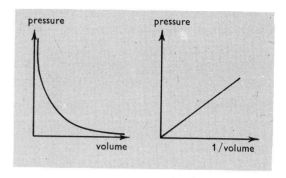

Fig. 119   Graphs for Boyle's law

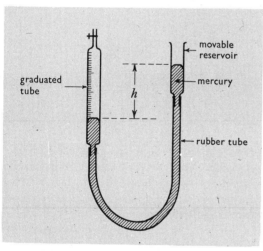

Fig. 118   Verification of Boyle's law

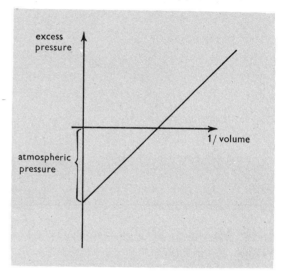

Fig. 120   Graph used with Boyle's law if atmospheric pressure unknown.

N.B. A little time should be allowed between altering the position of the levels and taking the reading so that if the gas has changed its temperature on expansion or contraction it may return to room temperature.

A graph of P against V is a hyperbola, whereas a graph of P against $1/V$ is a straight line. (N.B. P = total pressure; V = volume)

If atmospheric pressure is not known, a graph of excess pressure against $1/V$ may be plotted and a value for atmospheric pressure found.

22.3 Charles' law

The volume of a fixed mass of gas at constant pressure is directly proportional to the absolute temperature.

22.4 Experimental verification by the capillary tube method

The apparatus consists of a capillary tube sealed at one end and containing a short mercury thread which traps a fixed mass of air. A thermometer is strapped to the side of the tube and fulfils two functions:

(i)   To measure the temperature.

(ii) To measure the volume of the air column in terms of its length in thermometer divisions.

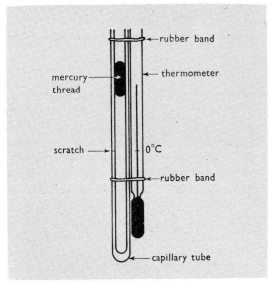

Fig. 121 Verification of Charles' law

The length from the bottom of the tube to the 0 mark on the thermometer, which is set against a scratch on the tube, must also be known in the same units. The apparatus is warmed in a water (etc.) bath, and corresponding readings of temperature and length taken, allowing the temperature to remain constant at any given value for some time so that the gas acquires that temperature. N.B. Pressure equals atmospheric plus that due to constant length of mercury thread throughout.

A graph is plotted of volume against temperature.

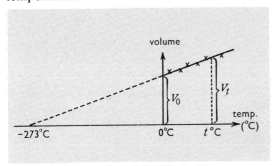

Fig. 122 Graph for Charles' law

Readings will only be obtained in the region marked, but the straight line so obtained may be extended back to cut the temperature axis at $-273°C$ approximately whatever gas is used. This does not represent the actual behaviour of the gas, which would liquefy.

The volume coefficient of expansion for the gas is defined as the increase in volume of unit volume at $0°C$ for each kelvin rise in temperature (α)

i.e. $$\alpha = \frac{V_t - V_0}{V_0 t}$$

and $$\alpha = \frac{1}{273} \quad \text{(approximately for all gases)}$$

Also by similar triangles

$$\frac{V_0}{273} = \frac{V_t}{273 + t}$$

Therefore if T_t is written for $273 + t$ and T_0 for 273

$$\frac{V_0}{T_0} = \frac{V_t}{T_t}$$

and a new scale of temperature, the absolute scale, with its zero at $-273°C$, and with its degrees the same size as Centigrade degrees can be introduced.

$$\left\{ \begin{array}{lr} \text{i.e. } °C \text{ to K (for Kelvin)} & \text{add } 273 \\ \text{K to } °C & \text{subtract } 273 \end{array} \right\}$$

N.B. $-273°C$ is often referred to as the absolute zero, an unobtainable minimum temperature.

22.5 The law of pressures

The pressure of a fixed mass of gas at constant volume is directly proportional to the absolute temperature.

22.6 The experimental verification by the constant volume gas thermometer

The gas under test is enclosed in the glass bulb which is connected by a capillary tube

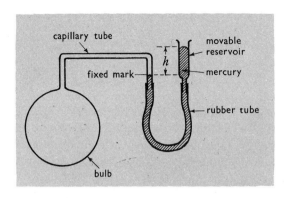

Fig. 123   Verification of the law of pressures

to a rubber tube containing mercury. The volume is maintained constant by adjusting the reservoir to bring the other mercury level back to the fixed mark. The bulb is heated to a known temperature, in a water (etc.) bath and when the gas has had sufficient time to reach this temperature the pressure required to maintain the chosen volume is recorded. The total pressure required will be $\pi + h$ where π is atmospheric pressure in m of mercury.

A graph of pressure against temperature is plotted.

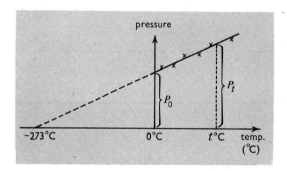

Fig. 124   Graph for the law of pressures

The graph is extended as before.

The pressure coefficient of expansion for the gas is defined as the increase in pressure, expressed as a fraction of the pressure at 0°C, for one kelvin rise in temperature (β)

$$\beta = \frac{P_t - P_0}{P_0 t} = \frac{1}{273} \quad \text{as before}$$

also the idea of absolute temperature may be used again

$$\frac{P_0}{T_0} = \frac{P_t}{T_t}$$

where

$$T_0 = 273$$
$$T_t = 273 + t$$

22.7   Use as a thermometer

The instrument can be modified into a standard thermometer. In this case the bulb is cylindrical in form and made of an alloy of platinum and iridium. The gas used is helium for low temperatures, hydrogen up to about 500°C, and above that temperature nitrogen (limit about 1500°C).

As atmospheric pressure is needed every time a reading is made, the standard instrument incorporates its own barometer, designed so that the total pressure is read off directly.

Disadvantages are the large size of the bulb, the 'dead space' of the capillary tube, and in the simple form the need to measure atmospheric pressure each time.

N.B. To interpret the results use

$$t = 100 \cdot \frac{P_t - P_0}{P_{100} - P_0}$$

when t is on the constant volume gas scale

22.8   The ideal gas equation

The above laws summarize the approximate behaviour of many gases for average laboratory conditions. Wide variations may occur under certain other conditions (see **§25.3**).

It is often useful to consider a gas which obeys these laws exactly. Such a gas is called an ideal or perfect gas. Consider such a gas originally at Pressure P_1, Volume V_1, and absolute temperature T_1. Let it change at constant temperature to P_2, V', T_1.

By Boyle's law

$$P_1 V_1 = P_2 V'$$

Now at constant pressure let it change to P_2, V_2, T_2.

By Charles' law

$$\frac{V'}{T_1} = \frac{V_2}{T_2}$$

$$\therefore \quad P_1 V_1 = \frac{P_2 V_2 T_1}{T_2}$$

or $\dfrac{P_1 V_1}{T_1} = \dfrac{P_2 V_2}{T_2} =$ a constant

and for 1 mole of this gas, the constant is defined as R the universal gas constant,

so $$PV = RT$$

Note that for 1 g of gas $PV = rT$ where $r = R/M$ ($M =$ molecular weight), and r, the gas constant per gram, is different for different gases.

23   CALORIMETRY

23.1   Units — Historical development

Basic experiments show that the amount of heat energy needed to warm water is proportional to both the mass of water and to the change in temperature. So, by choice of units,

Heat change = mass of water × change in temperature,

and the following units are used:

(*i*)   British Thermal Unit — the amount of heat required to raise the temperature of 1 lb of water through 1 F°

(*ii*)   Therm — equal to 10^5 Btu

(*iii*)   Calorie — the amount of heat required to raise the temperature of 1 g of water through 1°C.

The last unit is the only one of the three used in scientific work, yet is not a satisfactory definition for work of highest precision, so the next stage is to define the temperature change

e.g.   15° calorie, from 14·5°C to 15·5°C
mean calorie, 1/100 of the heat required for the change from 0°C to 100°C.

(These differ by 0·05%)

The final development is to measure heat energy directly in absolute mechanical units, joules, and this should now be standard practice.

23.2   Specific heat capacity

If materials other than water are used then it is found that different amounts of heat are needed to cause the same temperature change in the same masses of different substances. Therefore, a third factor, a constant for the material, called its specific heat capacity, enters into the equation

Heat change = mass × specific heat capacity × change in temperature.

The specific heat capacity of a material may be defined as the quantity of heat energy required to raise the temperature of unit mass of the substance through unit change of temperature.

The SI unit is J kg^{-1} K^{-1} (the former unit was cal g^{-1} °C^{-1}).

Thermal capacity

The quantity of heat energy required to cause unit temperature change for the whole object.

Water equivalent

The mass of water having the same thermal capacity as the object.

Both of these are numerically equal to mass × specific heat capacity, but thermal capacity has units joule K^{-1} (or cal °C^{-1}) and water equivalent has units kg (or g).

23.3   Heat and work

Early theories of the nature of heat looked at it in terms of a fluid, but experiments of

	Rumford (boring cannon barrels);
Davy	(rubbing ice in a vacuum);
and Joule	(showing amount of heat produced depended only on work done and not on the procedure);

were evidence in favour of heat as a form of energy.

Joule's 'paddle wheel' and other experiments led to the equation

$$\frac{\text{Work done}}{\text{Heat produced}} = J$$

where J is the mechanical equivalent of heat.

Now that heat energy is measured in joule this conversion equation is not needed, but J.P. Joule's work shows that specific heat capacity is a true constant for the material and leads to one method of measuring specific heat capacity.

23.4   The determination of the specific heat capacities of solids.

(a) Callendar's method

The heat generated by friction between the band and the copper, causes a rise in temperature of the copper drum.

The handle is turned for a known number of revolutions in a known time, trying to keep the spring balance reading constant. A cooling curve is plotted for the drum at the end.

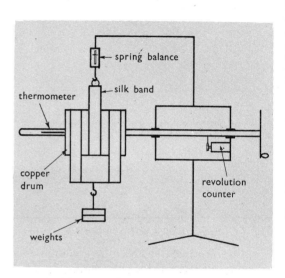

Fig. 125   Callendar's apparatus

Work done = Frictional force × Circumference of drum × Number of revolutions
$$= (W - w) \times 2\pi r \times n$$

where W = weight applied
w = spring balance reading
r = radius of drum
n = number of revs

Heat produced = Mass of drum × Sp. ht. capacity × Rise in temperature
$$= mc\theta$$

where m = mass of drum
c = sp. ht. capacity of drum
θ = temperature rise

so $\qquad (W-w)2\pi rn = mc\theta$
whence c can be found

As the heating goes on for some time a cooling correction is needed. So plot a graph of temperature against time.

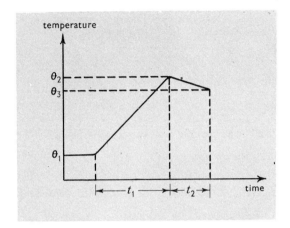

Fig. 126   Heating curve for electrical calorimetry

At θ_2 the rate of loss of temperature is

$$\frac{\theta_2 - \theta_3}{t_2}$$

At θ_1 the rate of loss of temperature is 0
So the average rate of loss of temperature is

$$\frac{\theta_2 - \theta_3}{2t_2}$$

as the average excess temperature is half the maximum excess. This assumes Newton's law of cooling, that the rate of loss of heat is proportional to the excess temperature, and that θ_1 is room temperature.

So the true temperature change produced is

$$(\theta_2 - \theta_1) + \frac{(\theta_2 - \theta_3)t_1}{2t_2}$$

(b)   *Method of mixtures*

This depends on two principles:

(i)   That if hot and cold objects are placed in thermal contact, there will be a heat flow to obtain a temperature equilibrium.

(ii)   The heat lost by the hot objects equals the heat gained by the colder ones, if the heat lost from the whole system may be ignored.

Usually elementary experiments are performed by dropping a hot object into water in a calorimeter and measuring:

Mass of calorimeter
Mass of calorimeter + water $\therefore$ mass of water
Mass of calorimeter + water + solid $\therefore$ mass of solid
Temperature of hot solid
Temperature of cold water
Temperature of mixture

Then assuming the specific heat capacity of the calorimeter, the calculation may be made by equating the heat lost by a hot object to the heat gained by calorimeter plus the heat gained by the water.

Precautions

(i)   The specimen must be uniformly at the higher temperature.

(ii)   There must be a quick transference without splashing.

(iii)   The mixture should be continuously stirred, allowing for thermal capacity of stirrer.

(iv)   Reduce heat losses by lagging, shielding and polishing the calorimeter and using a lid.

(v)   Apply a cooling correction, e.g. Rumford's (start with the calorimeter and contents as much below room temperature as it is expected to go above).

(c)   *Electrical heating*

A metal block is drilled to take a special

heater and a thermometer. The electrical circuit is as shown.

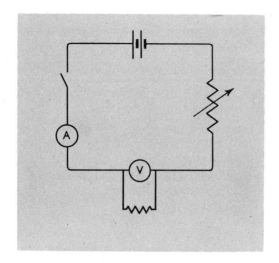

Fig. 127 Electrical circuit for calorimetry

Then the amount of heat produced by a current (I) flowing under a p.d. (V) for a time (t) is IVt (joule) and this is equal to the mass of the block multiplied by its specific heat capacity and its rise in temperature.

A cooling correction is needed as in (a)

(d) *Nernst and Lindemann's vacuum calorimeter*

For solids, this method can be used over wide ranges of temperature, particularly at low temperatures. Any one observation is made over a small change in temperature.

The specimen is machined to the shape shown, and an insulated platinum coil wound on it. This coil acts as both heater and resistance thermometer.

The specimen is brought to just below the temperature at which the measurement is to be made, and its enclosure evacuated. Its temperature is measured, a known amount of electrical energy supplied, and the final temperature found.

After correcting for radiation and the thermal capacity of the coil, an accurate result is calculated.

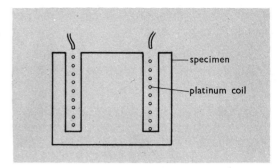

Fig. 128 Specimen for Nernst and Lindemann's calorimeter

23.5 The specific heat capacity of liquids

(a) *Method of mixtures*

(i) Use a solid of known specific heat capacity which does not react with the liquid.

(ii) Use two liquids and allow them to mix thermally but not physically.

(b) *Method of cooling*

The method is based on the idea that if two externally identical bodies cool in the same

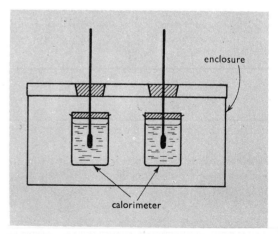

Fig. 129 Specific heat capacity of a liquid by cooling

enclosure the rates of loss of heat at a given temperature will be the same. This is independent of any particular law of cooling.

Two small calorimeters, as identical as possible are filled with equal volumes of two liquids whose specific heat capacities are to be compared.

They are both heated in hot water then transferred to the enclosure (ideally in a larger outer vessel of cold water). Cooling curves are plotted for both, covering at least one common range of 20K to 30K.

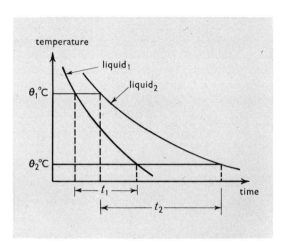

Fig. 130   Cooling curves for two liquids

Let the mass of the first calorimeter be M_1, its specific heat capacity c, the weight of liquid in it m_1, and its specific heat c_1. From the graph it cools from $\theta_1°C$ to $\theta_2°C$ in t_1 seconds.

Therefore the average rate of loss of heat is

$$\frac{(M_1c + m_1c_1)(\theta_1 - \theta_2)}{t_1}$$

and similarly for the second calorimeter and liquid.

Now these two rates are equal, so

$$\frac{(M_1c + m_1c_1)(\theta_1 - \theta_2)}{t_1} = \frac{(M_2c + m_2c_2)(\theta_1 - \theta_2)}{t_2}$$

and from this c_1 can be found if c_2 is known.

N.B.

(*i*) m_1 and m_2 are found at the end of the experiment.

(*ii*)   Another method of comparison is to draw tangents at a particular temperature and use the respective gradients, but there is not much difference in practice.

(c)   Electrical heating

The procedure is the same as in §23.4(c) but a suitable calorimeter and heating coil is needed.

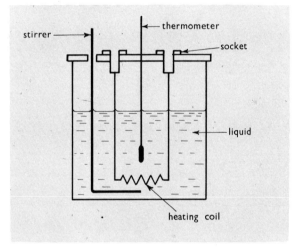

Fig. 131   Calorimeter with heating coil

(d)   Continuous flow calorimetry (Callendar and Barnes)

A steady flow of liquid from a constant head apparatus passes through the tube and is heated by the length of resistance wire. The electrical circuit is the same as in (c) above. Heat losses are minimized by the vacuum jacket.

The arrangement is set up and allowed to stand until the temperatures θ_1, θ_2 and the other quantities are constant.

If mass m of liquid of specific heat capacity c, flow per second

$$IV = mc(\theta_2 - \theta_1) + H_0$$

F

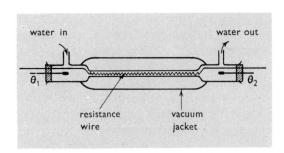

Fig. 132   Callendar and Barnes's continuous flow
calorimeter

where H_0 is the heat lost per second. The
rate of flow and the electrical current are then
altered to give the same values of θ_1 and θ_2,
but new m', I' and V'. As the temperatures
are the same H_0 is assumed to be the same
also. Thus:

$$I'V' = m'c(\theta_2 - \theta_1) + H_0$$

subtracting

$$IV - I'V' = (m - m')c(\theta_2 - \theta_1)$$

hence c can be found.

The advantages of this method are:

(*i*)   All the readings are made under
steady state conditions.

(*ii*)   The water equivalent of the apparatus
is not required.

(*iii*)   Errors due to heat losses can be
eliminated.

(*iv*)   Variation of specific heat capacity
with temperature can be investigated.

(*v*)   All readings (apart from m) can be
made electrically, (using platinum resistance
thermometers) and are therefore capable of
great accuracy.

The disadvantages are:

(*i*)   A large quantity of liquid is required.

(*ii*)   The theory assumes constant surrounding conditions over a long time.

(*iii*)   The input temperature of the liquid
may vary, particularly if tap water is being
used.

(*iv*)   Electrolysis can occur.

23.6   The specific heat capacities of gases

The specific heat capacity of a gas depends
on the conditions under which it is measured,
for if the gas is allowed to expand it will do
work, and so absorb more heat energy. Theoretically an infinite number of specific heat
capacities could be defined, but in practice
only two are commonly met with; when the
gas is heated at constant pressure or constant
volume.

The specific heat capacity at constant
pressure (c_p) is the amount of heat required to
raise the temperature of 1 kg of the gas through
1 K when the pressure is maintained constant.

The specific heat capacity at constant
volume (c_v) is the amount of heat required to
raise the temperature of 1 kg of the gas
through 1 K when the volume is maintained
constant.

Both are usually measured at atmospheric
pressure. c_p is greater than c_v, and the ratio
of the principal specific heats of a gas $c_p/c_v = \gamma$
is of considerable significance (see §**23.14**).

23.7   To show that $c_p - c_v = r$

Imagine a cylinder with a frictionless piston
containing one mole (i.e. mass M) of gas, at a
pressure P.

At constant volume, for a change of 1 K

$$\text{Heat supplied} = Mc_v$$

At constant pressure, for a change of 1 K

$$\text{Heat supplied} = Mc_p$$

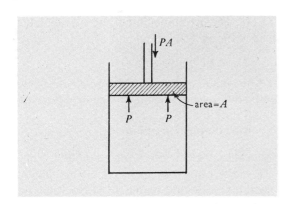

Fig. 133   Relation between the principal specific heats of a gas

or, it is equal to the heat supplied at constant volume, plus the heat energy required to push the piston back a distance x to restore constant pressure.

The force on the piston = Pressure × Area
$$= P \cdot A$$

∴   Work done in moving $x = PAx$

∴ $Mc_p = Mc_v + PAx$

$$= Mc_v + P.\delta V$$

where V is volume of the gas, and δV the change in volume.

but $PV = RT$

and if a change of temperature δT causes a change of volume δV at constant pressure

so $P(V + \delta V) = R(T + \delta T)$

∴ $P\delta V = R\delta T$

but in this case $\delta T = 1$

∴ $P\delta V = R$

so $Mc_p = Mc_v + R$

and dividing by M

$$c_p - c_v = r \quad \text{where } r = R/M$$

23.8   The determination of the specific heat capacities of gases

(a)   *Joly's differential steam calorimeter* (c$_v$)

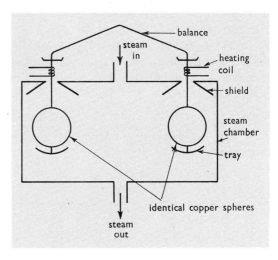

Fig. 134   Joly's differential steam calorimeter

The two identical hollow copper spheres are evacuated and one filled under pressure with the gas whose specific heat is being measured. The balance is counterpoised and steam is passed until the whole of the inside of the steam chamber has reached the temperature of the steam. More steam will condense on the sphere containing the gas, and the extra weight may be found from the change in counterpoising weight on the balance.

This weight multiplied by the specific latent heat of vaporization of water equals the weight of the gas multiplied by its specific heat capacity (c_v) and its rise in temperature.

The trays retain the condensed steam, the shields prevent drops of water condensed on the walls of the steam chamber from falling onto the spheres, and the heating coils keep the suspending wires free of water drops which might impede the swing of the balance.

Corrections must be made for the slight expansion of the gas and for the inevitable slight differences between the spheres.

(b)   Continuous flow method (c_p)

The method of Callendar and Barnes has been adapted for use with gases. The gas is supplied from a cylinder maintained at constant temperature and circulates through the apparatus and past the heating coil at constant pressure. The same measurements are made as in the case of a liquid, but the mass of gas flowing is found from the volume, temperature and fall in pressure of the gas in the cylinder.

23.9   The temperature variation of specific heat capacity

The changes in specific heat capacity are evidence for the quantum theory.

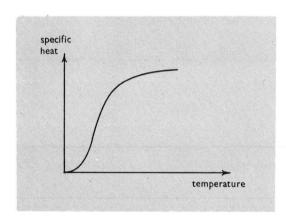

Fig. 135   Variation of specific heat capacity with temperature

23.10   The specific latent heat of fusion

The specific latent heat of fusion of a substance is the quantity of heat required to convert unit mass of solid to liquid without change in temperature.

(Units J kg^{-1} or cal g^{-1});
Equation, Heat change = Mass × specific latent heat)

23.11   The determination of the specific latent heat of fusion

(a)   Method of mixtures

Use small pieces of solid (e.g. dried ice) and add, one at a time, to calorimeter and liquid until temperature has fallen sufficiently. Apply a cooling correction.

(b)   Nernst and Lindemann — modified

Use the same principle as described above, but the specific heat capacity of the solid, the specific latent heat of fusion, and the specific heat capacity of the liquid will be involved.

23.12   The specific latent heat of vaporization

The specific latent heat of vaporization of a substance is the quantity of heat required to convert unit mass of liquid to vapour without change in temperature.

23.13   The determination of the specific latent heat of vaporization

(a)   Method of mixtures

Vapour is condensed in its own liquid. The main trouble is to free vapour of liquid mixed with it, either by 'boiling over' or by partial condensation.

One arrangement which minimizes this is shown in the diagram.

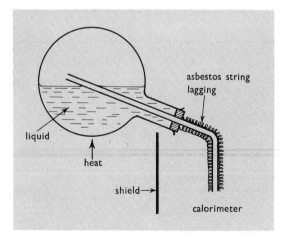

Fig. 136   Vapour supply for latent heat calorimetry

(b) Electrical method

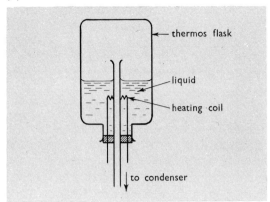

Fig. 137 Electrical determination of latent heat of vaporization

The liquid is boiled electrically and the weight condensing in a given time is found.

As the heat supplied can be found electrically the specific latent heat can be quickly calculated.

The thermos flask minimizes heat losses from the hot liquid.

23.14   Isothermals and adiabatics

There are two theoretical sets of conditions under which compression and expansion of gases can occur which are of great importance.

(a) Isothermal

The change occurs at constant temperature, (and therefore in practice slowly). Boyle's law is obeyed, and $PV = $ constant.

(b) Adiabatic

The change occurs so that no heat can enter or leave the gas during the change. This usually means a quick change. Boyle's law is not obeyed as the temperature changes.

$$PV^\gamma = \text{constant},$$

where $\qquad \gamma = c_p/c_v.$

The equation for an adiabatic will not be proved at this stage.

N.B.

(i) Both equations are true only for an ideal gas.

(ii) The equation $PV = RT$ is true in either case, and combining this with the second equation

$$PV^\gamma = (PV) \cdot V^{\gamma-1} = \text{constant}$$

$$\therefore \qquad RT \cdot V^{\gamma-1} = \text{constant}$$

or $\qquad TV^{\gamma-1} \qquad = \text{constant}'$

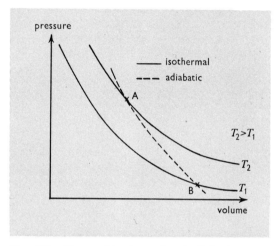

Fig. 138 Slope of isothermal and adiabatic curves

Consider gas at A, expanded adiabatically to B. As the temperature falls, B must be on a lower isothermal, hence adiabatic is steeper than an isothermal.

23.15   The work done in expansion

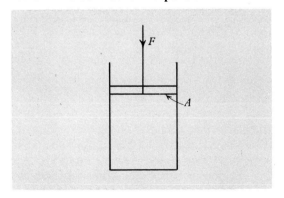

Fig. 139 Work done by a gas in expanding

Consider some gas contained in a cylinder of cross-sectional area A by a frictionless piston. The force on the piston equals F and if the piston moves a small distance x

$$\text{Work done} = \text{Force times Distance}$$
$$= Fx$$
$$= \frac{F}{A} \cdot Ax$$

$$= P \cdot \delta V$$
$$P = \text{gas pressure}$$
$$\delta V = \text{Change in volume}$$

So the work is measured by the area under a P–V curve or, if a loop is formed, it is measured by the area of the loop.

24 VAPOURS

24.1 Saturated and unsaturated vapours

If a liquid is placed in a closed vessel an extra pressure will be observed, due to the bombardment of the vessel walls by the molecules which have escaped from the liquid.

This pressure, the vapour pressure, rises until a dynamic equilibrium is reached, and the number of molecules leaving the liquid equals the number returning. The maximum pressure is called the saturation vapour pressure (s.v.p.).

Before equilibrium, when more substance could go into the vapour state, it is said to be an unsaturated vapour. A saturated vapour exists when no more substance, in total, can go into the vapour state, and is usually identified by the presence of excess liquid.

If a small quantity of liquid is inserted into a barometer the mercury falls, due to the vapour pressure of the unsaturated vapour. If more liquid is added, the mercury falls to a definite level, and excess liquid appears on the mercury surface. The pressure indicated is the s.v.p. of the substance concerned.

If the barometer tube is heated the s.v.p. increases, and the mercury falls further. At the boiling point of the liquid the mercury is completely depressed, so that the boiling point of a liquid is the temperature at which the s.v.p. equals the external pressure, and vapour can form inside the liquid.

24.2 Vapour pressure graphs

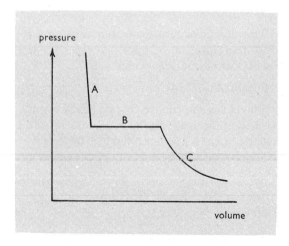

Fig. 140   Pressure–volume curve for a vapour

The graph applies to a fixed mass of substance at a fixed temperature.

At A only liquid is present, and it is difficult to compress.

At B liquid and saturated vapour are present, and the pressure is independent of the volume.

At C unsaturated vapour is present, and Boyle's law is approximately obeyed.

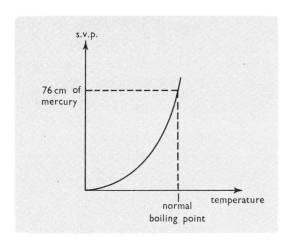

Fig. 141   Variation of s.v.p. with temperature

Enough substance must be present to keep the vapour saturated.

24.3   The determination of s.v.p.

(a)   *Barometer method*

As above.

(b)   *Ramsay–Young*

The liquid is boiled at various reduced pressures. In equilibrium the s.v.p. equals the air pressure within the apparatus, and so the s.v.p. at various temperatures may be found.

(c)   *Tube method*

This depends on Dalton's law of partial pressures: The pressure exerted by a mixture of gases and vapours is the sum of the pressures they would exert separately if each alone occupied the same total volume at the same temperature. This is valid only if each gas or vapour does not react with any other.

A tube containing an index of the liquid under test is strapped to a thermometer so that they can be heated in a liquid bath. The length (and therefore volume) of the

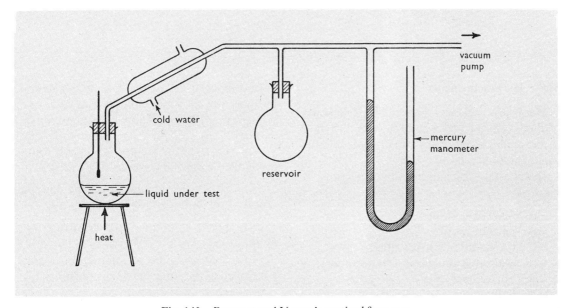

Fig. 142   Ramsay and Young's method for s.v.p.

air–vapour column is measured in thermometer divisions at different temperatures.

Let π = atmospheric pressure

S_1 = s.v.p. at temperature T_1 (absolute)

S_2 = s.v.p. at temperature T_2

L_1 = length at temperature T_1

L_2 = length at temperature T_2

Then partial pressure of air at $T_1 = \pi - S_1$, at $T_2 = \pi - S_2$ so apply ideal gas equation (to the air)

$$\frac{(\pi - S_1)L_1}{T_1} = \frac{(\pi - S_2)L_2}{T_2}$$

and if S_1 is known at one temperature, it may be found at others.

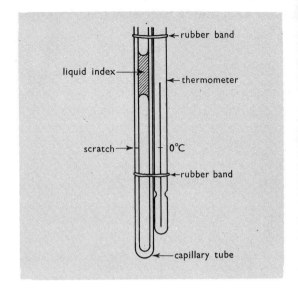

Fig. 143 Tube method for s.v.p.

25 THE KINETIC THEORY OF GASES

25.1 The kinetic theory

Many properties of materials, particularly of fluids, can be explained by assuming them to be made up of many small molecules in rapid, random motion.

Qualitative explanations should already be known for:

(i) Brownian motion,

(ii) Surface tension.

(iii) Evaporation (and cooling).

(iv) Saturated and unsaturated vapours.

(v) Latent heat.

(vi) Diffusion.

(vii) Pressure of a gas.

In the case of gases, a simple algebraic treatment of the kinetic theory can be worked out, starting with the following assumptions:

(i) That the molecules are small, round, perfectly elastic and far apart.

(ii) That the molecules are moving rapidly, and therefore collide frequently.

(iii) That the number of molecules is very large.

(iv) That the time of interaction is small compared to the time between collisions.

(v) That the mutual attraction of the molecules may be ignored.

25.2   The pressure due to a gas

Consider a molecule of gas in a cube of side l. Let the molecule have mass m, and velocity c, resolvable into three components u, v, w, along axes perpendicular to the sides of the cube.

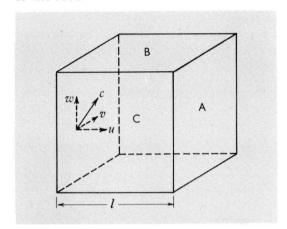

Fig. 144   Kinetic theory equation for pressure

Consider the u component.
As the molecule is perfectly elastic, velocity after collision is $-u$.

$$\therefore \quad \text{Change of velocity} = 2u$$

$$\text{Change of momentum} = 2mu$$

$$\text{Time between collisions at A} = \frac{2l}{u}$$

$\therefore$ Number of collisions per second $= u/2l$
$\therefore$ Average force on A due to this molecule = momentum change per collision $\times$ number of collisions per sec

$$= 2mu \times \frac{u}{2l} = \frac{mu^2}{l}$$

$$\therefore \quad \text{Pressure on A} = \frac{\text{Force}}{\text{Area}} = \frac{mu^2}{l^3}$$

If there are N molecules

$$\text{Total pressure on A} = P_A = \frac{Nmu^2}{l^3}$$

$$\text{or more strictly} \quad \frac{m}{l^3} \sum_{1}^{N} u^2$$

Similarly for other faces

$$P_B = \frac{m}{l^3} \sum_{1}^{N} w^2 \quad P_C = \frac{m}{l^3} \sum_{1}^{N} v^2$$

If cube is small $P_A = P_B = P_C$

so
$$3P = P_A + P_B + P_C$$

$$= \frac{m}{l^3} \sum_{1}^{N} (u^2 + v^2 + w^2)$$

$$= \frac{m}{l^3} \sum_{1}^{N} c^2$$

so
$$3P = \frac{m}{V} \sum_{1}^{N} c^2$$

where V = volume = l^3

now
$$\sum_{1}^{N} c^2 \equiv N\overline{c^2}$$

where $\overline{c^2}$ is mean square velocity

so
$$3P = \frac{Nm\overline{c^2}}{V}$$

or
$$P = \tfrac{1}{3}\rho\overline{c^2}$$

where ρ = gas density = Nm/V

Also the mean K.E. of the molecules = $\tfrac{1}{2}m\overline{c^2}$ and this can be shown to be proportional to the absolute temperature, and to be the same for all gases at the same temperature.
N.B.

(i)   At constant T $m\overline{c^2}$ is constant

$$\therefore \quad \frac{Nm\overline{c^2}}{3} \text{ is constant}$$

i.e. $PV = \text{constant}$ (Boyle's law)

(ii) $PV = \dfrac{Nm\overline{c^2}}{3} = \dfrac{2N}{3} \cdot \dfrac{m\overline{c^2}}{2} = \text{constant} \times T$

(ideal gas equation)

25.3   Real gases

Thus simple kinetic theory agrees with the ideal gas equation. However, experiments with real gases show important deviations from the simple equations.

There are two important sets of results.

(a) Andrews' experiments

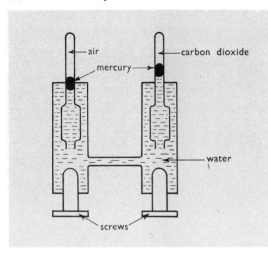

Fig. 145   Andrews' apparatus

By turning the screws different pressures can be set up. The pressure is measured by

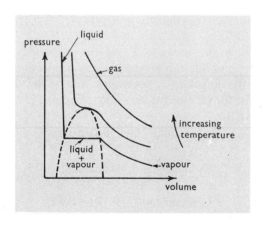

Fig. 146   Isothermals from Andrews' experiments

the air column and the corresponding carbon dioxide volume noted.

Note.

(i) There is a temperature above which no liquefaction occurs, however high the pressure. This is called the critical temperature.

(ii) The distinction between gas and vapour, i.e. a gas exists only above the critical temperature.

(b) Amagat's experiments

Amagat measured the product PV for a wide range of pressures up to 400 atmospheres. The resulting graphs showed temperature dependence.

T_B = Boyle's temperature, i.e. temperature at which Boyle's law is exactly obeyed.

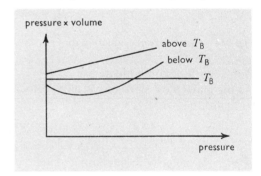

Fig. 147   Results of Amagat's experiments

These deviations occur mainly because the volume of the molecules and their mutual attraction is ignored. Many equations exist which try to bring these into account,
e.g.   Van der Waal's equation

$$\left(P + \frac{a}{V^2}\right)(V - b) = RT$$

where a and b are extra constants.

26 THERMAL CONDUCTIVITY

26.1 The basic ideas

If one end of a metal rod is heated, heat will flow down the rod. If the temperature can be measured at different points along the bar the the following observations can be made:

(*i*) The temperatures will reach equilibrium values, i.e. a steady state has been reached, with respect to time.

(*ii*) The temperature distribution depends on the degree of lagging:

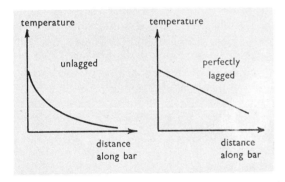

Fig. 148 Temperature distribution along a metal rod heated at one end

In the unlagged case more heat is lost from the first hot part of the sides and so the temperature falls more rapidly. When perfectly lagged no heat is lost from the sides, and so all the heat flows down the bar.

It is this last state that is usually concerned.

It can be realized by either perfect lagging or by imagining the central part of an infinite slab.

Heat flow is found to be proportional to:

(*i*) Area perpendicular to heat flow.
(*ii*) Temperature difference.
(*iii*) Time.

(*iv*) The reciprocal of thickness.

Thus
$$Q \propto \frac{A(\theta_2 - \theta_1)t}{d}$$

where A = cross-sectional area,
d = thickness
t = time
θ_2, θ_1 = temperatures of faces

and
$$\frac{Q}{t} = \frac{\lambda A(\theta_2 - \theta_1)}{d}$$

where λ is the coefficient of thermal conductivity, defined as the quantity of heat flowing across unit area of an infinite slab in unit time under unit temperature gradient when steady state has been reached.

$$\left(\text{Temperature gradient} = \frac{\theta_2 - \theta_1}{d} \right)$$

Units: $W\,m^{-1}\,K^{-1}$ (or $cal\,cm^{-1}\,s^{-1}\,C^{\circ-1}$)

26.2 The determination for a good conductor (Searle's method)

The metal specimen, in the form of a rod, is heated at one end, and being well lagged, all the heat flows to the other end and is absorbed by cold water flowing round in a copper tube.

The apparatus is left to attain steady state and then the four temperatures and the mass (m) of cooling water flowing per s are measured. The area of the rod (A) and the distance (d) between the thermometers in the specimen is also determined.

$$\text{Heat flowing/s} = \frac{\lambda A(\theta_1 - \theta_2)}{d} = mc(\theta_3 - \theta_4)$$

where c = sp. ht. capacity of water

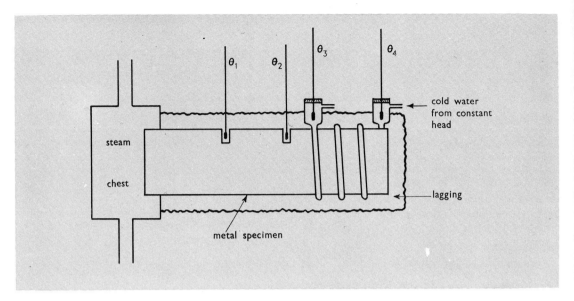

Fig. 149   Searle's apparatus

26.3   The determination for a bad conductor (Lee's disc)

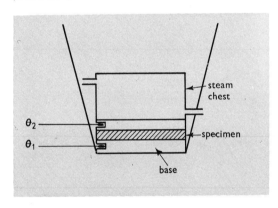

Fig. 150   Lee's disc

The specimen, in the form of a disc, is sandwiched between a steam chest and the base, into which thermometers can be placed.

The whole is suspended until a steady state is reached. Then the specimen is removed, so that the base warms up a little above its equilibrium temperature, the steam chest is then removed, the specimen replaced and a cooling curve is plotted.

It is assumed that the heat loss from the base is the same at the same value of θ_1 in both cases. In the first case it must equal the heat flowing by conduction through the specimen, and in the second case to mass × sp. ht. cap. × rate of fall of temperature of the base.

Thus $$\frac{dQ}{dt} = \frac{\lambda A (\theta_2 - \theta_1)}{d} = mc\frac{d\theta}{dt}$$

where dQ/dt = heat lost/s
      λ = coefficient of thermal conductivity
      A = area of specimen
      d = thickness of specimen
      m = mass of base
      c = sp. ht. capacity of base
      $d\theta/dt$ = rate of fall of temperature of base

$d\theta/dt$ must be measured from the graph by drawing a tangent at θ_1.

26.4   Sandwich problems

Where conduction takes place through several layers in 'sandwich' form, the interfacial temperatures must be introduced and the heat flow through each layer is equal.

27   CONVECTION

A full theory of convection is too difficult to consider. However, one special case must be considered.

27.1   Newton's law of cooling

The rate of loss of heat is proportional to the temperature excess of the body over its surroundings.

The law was first discovered by Newton, who confined its validity to cases of forced convection. In general it seems to apply over moderate temperature differences (possibly 200 K) for forced convection (i.e. where heat loss is mainly by convection), and over small temperature differences for natural convection (more strictly a $\frac{5}{4}$ power law applies).

To test its validity, a cooling curve is plotted for a calorimeter containing water placed in a draught on a cork block. Tangents are drawn to the graph at various points, and a second graph of rate of loss of temperature ($\propto$ heat) against excess temperature is plotted.

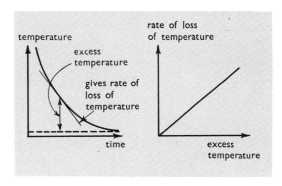

Fig. 151   Newton's law of cooling

28   RADIATION

This is the third method of heat propagation, distinguished by its speed and ability to pass through a vacuum; in general its properties are very similar to those of light.

28.1   Detectors of radiation

Various detectors, some very sensitive, exist, but only two need be mentioned. Both respond to the heat formed when the radiation is absorbed.

(*a*)   *Differential air thermometer*

As the blackened bulb absorbs more heat, the gas inside it warms up more, expands more and forces the liquid round.

(*b*)   *Thermopile*

The ends of the thermo-couples exposed to the radiation get hotter, setting up a (small) e.m.f. suitable for detection by a sensitive galvanometer.

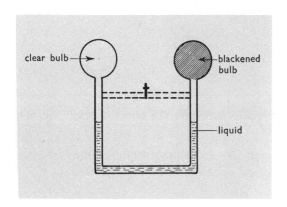

Fig. 152   Differential air thermometer

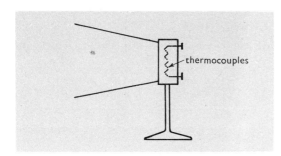

Fig. 153   Thermopile

28.2   The absorption and emission of radiation

If a thermopile is placed at equal distances from various surfaces at the same temperature (e.g. Leslie's cube) it is found that a matt black surface is the best emitter.

The differential air thermometer shows directly that black surfaces are also the best absorbers.

28.3   A black body

This is defined as one that will absorb all radiation falling on it.

Lampblack is about 96% efficient.

A wedge or sphere with a narrow aperture about 97% efficient.

28.4   Prevost's theory of exchanges

Consider an object A at temperature T_1 in an enclosure B, at temperature T_2. Let $T_1 > T_2$.

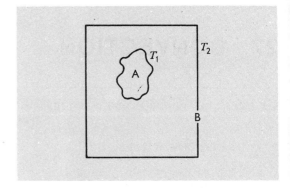

Fig. 154   Prevost's theory of exchanges

Prevost's theory states that both A and B will emit and absorb radiation, but A emits more than it absorbs so its temperature falls.

When $T_1 = T_2$, A emits and receives equal amounts of radiation, a dynamic equilibrium is established. As A does not alter the radiation distribution it cannot be 'seen' inside B.

If T_2 is then lowered, A will then receive less radiation, and so its temperature falls.

28.5   The law for the loss of heat by radiation (from a black body)

Dulong and Petit first investigated this by observing the rate of cooling of a blackened thermometer bulb placed in an evacuated enclosure.

Their results were analysed by Stefan, who showed that the heat radiated was proportional to the fourth power of the absolute temperature.

$$\therefore \quad E = \sigma(T_1^4 - T_0^4)$$

where E = energy lost per unit area per second.

T_1 = temperature of object (absolute)

T_0 = temperature of surroundings (absolute)

σ = Stefan's constant.

N.B. E can be in Wm^{-2} (or $cal\ cm^{-2}\ s^{-1}$).

28.6   The determination of Stefan's constant

A blackened copper sphere is heated electrically in an evacuated, water-jacketed enclosure, and the temperature observed for different power dissipations.

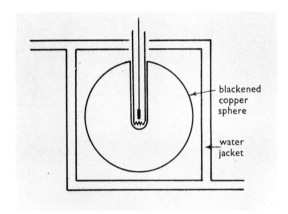

Fig. 155   Determination of Stefan's constant

$$IV = 4\pi r^2 \sigma (T_1{}^4 - T_2{}^4)$$

where I = current

V = p.d.

r = radius of sphere

σ = Stefan's constant

T_1 = temperature of sphere (K)

T_2 = temperature of water (K)

28.7   Emissivity

As a black body cannot be exactly realized in practice, these equations are modified by bringing in a multiplying constant (< 1) called the emissivity of the surface.

28.8   Pyrometers

The temperature of a hot object may be measured in terms of the radiation emitted by it. Instruments which enable this to be done are called pyrometers.

28.9   The distribution of energy in the spectrum

If the energy emitted by a black body at different temperatures is analysed by wavelength, very important results are obtained.

E_λ = energy emitted at wavelength λ

λ = wavelength of radiation

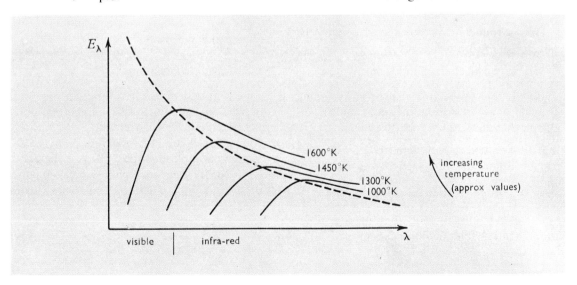

Fig. 156   Distribution of energy in black body spectra

Note

(*i*)   General increase in E as temperature is raised.

(*ii*)   There is a definite maximum, moving to lower wavelengths at higher temperature (cf. colour of radiation).

(*iii*) $\lambda_{max.}\ T = \text{constant}$ (Wien's displacement law).

(*iv*)   Classical physics cannot explain the shape of these curves, and quantum ideas must be used.

28.10   The electromagnetic spectrum

All the radiations which are part of the electromagnetic spectrum will cause heating when they are absorbed.

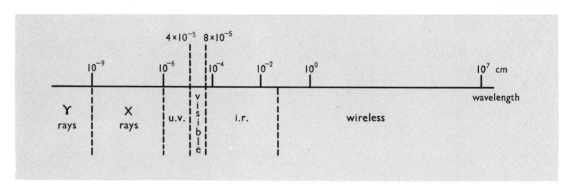

Fig. 157   The electromagnetic spectrum

The boundaries are very vague.

28.11   Ultra-violet radiation

This is found in certain spectra, and it is detected in the following ways:

(*i*)   It can blacken photographic plates.

(*ii*)   It can cause fluorescence, for example, uranium oxide or barium platinocyanide emit green light when exposed to ultra-violet.

N.B. The wavelength emitted is always greater than the wavelength received in fluorescence.

(*iii*)   It can cause phosphorescence, for example, calcium sulphide continues to glow after the ultra-violet radiation is removed.

(*iv*)   It can cause photoelectric emission (see §**49.1**)

Typical uses of ultraviolet radiation are in medicine and microscopy.

28.12   Infra-red radiation

This is found in black body emission, and can be detected either by its photographic effect or by radiation instruments.

Typical uses of infra-red radiation are in medicine and photography. It is especially suitable for long distance photography in poor conditions as it is scattered much less than visible light, by the atmosphere.

WAVE THEORY

29 VIBRATIONS

29.1 Free vibrations

If any source of sound is set into vibration and then not interfered with by any external means, it will vibrate with its own natural frequency determined by the appropriate elastic forces within it. This will be of constant pitch, but diminishing in loudness. In some cases this frequency can be calculated. Such vibrations are said to be free.

The vibrations die away because of damping. A tuning fork is lightly damped as the vibrations persist for a long time. A bottle which emits a note when it is blown across the top, is heavily damped as the vibrations die away very rapidly.

29.2 Forced vibrations

When a vibrating tuning fork is held over the neck of an empty bottle, a feeble sound of the same frequency as the fork is heard from the air in the bottle. This is known as forced vibration. The air in the bottle is forced to vibrate with the same frequency as the fork, but the resultant amplitude is small. The body of a violin, the sounding board of a piano, a tuning fork placed on a table, are other examples of this type of vibration.

29.3 Resonance

If the frequency of the source causing forced vibrations can be altered, it is possible that it may become the same as the frequency of the free vibrations of the system. In this case the response of the system is maximum, and may be very large indeed. Resonance is said to occur; e.g. parts of some cars will vibrate noticeably at certain speeds. The engine frequency matches the natural frequency of vibration and causes resonance.

29.4 Barton's pendulums

This experiment illustrates the above ideas.

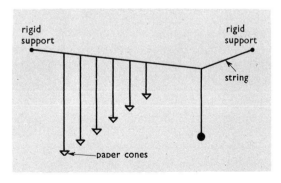

Fig. 158 Barton's pendulums

G

Several simple pendulums with bobs of paper cones are suspended from a string. A heavy bob is also suspended, as shown in the diagram.

The paper cones can be seen to be heavily damped. When the heavy pendulum is set swinging the lighter ones are set into forced vibration, but there is little selective resonance.

If the paper cones are loaded, they are much more lightly damped, and the resonance is easily seen.

The concept of resonance will be met again in alternating current circuits.

(*i*) Unloaded paper cones, heavily damped, unselective resonance.

(*ii*) Loaded pendulums, lightly damped, selective resonance.

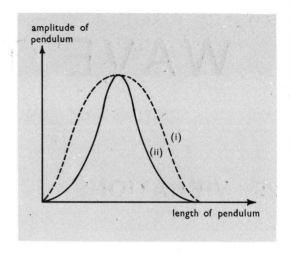

Fig. 159 Resonance curves for Barton's pendulums

30   WAVE MOTION. WAVES ON STRINGS

30.1   Wave-motion

With any type of wave motion the following features can be distinguished.

(*i*)   It is repetitive.

(*ii*)   There is no overall displacement of any particle, only an oscillation about a mean position.

(*iii*)   It is a means of transmitting energy from one place to another.

(*iv*)   It can be related, often very simply, to simple harmonic motion and therefore to a sine curve.

Without specifying any particular type of wave, but using the sine curve, certain important terms can be defined.

Wavelength λ

The distance from crest to crest, or any two corresponding points on the wave.

Velocity v

The rate at which the outline of the wave is travelling in the direction of the wave.

Amplitude a

The maximum displacement.

Frequency f or v

The number of complete waves passing a

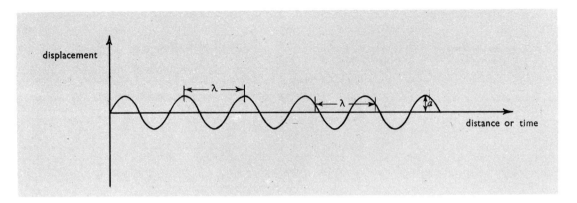

Fig. 160   Definition of wavelength and amplitude

given point per second, and the frequency of oscillation of any chosen particle.

Periodic time T

The time for one wave to pass a given point

$$\therefore \quad T = \frac{1}{f}$$

N.B. In one second f waves each of length λ pass a given point $\therefore$ the wave outline travels $f\lambda$ in one second.

$$\therefore \quad v = f\lambda$$

30.2   The types of waves

There are many different types of wave motion which fit the above pattern.

(a) Water waves

These can be observed at sea or when a stone is thrown into a pond. A bobbing cork shows that there is no overall movement of the water.

Controlled water waves, formed in a ripple tank enable various properties of the waves to be examined.

The connection with a sine curve is obvious by direct examination.

(b) Sound waves

These are caused by an object vibrating within the limits of about 20–15,000 Hz. They can be detected by the ear. Removing

the medium (e.g. air) stops the wave propagation.

Replacing the ear by a microphone coupled to a cathode ray oscilloscope shows that pure sound waves are sinusoidal.

N.B. Waves of this type exist above the audible limit but are called ultrasonic.

(c) Electromagnetic waves

This large family of waves, including radio, light, infra-red and ultra-violet etc. have similar properties.

Generators and detectors vary with the type of wave, and no material medium is needed for their propagation.

Again it is possible to represent the wave in terms of a sine curve.

30.3   The types of wave-motion

(a) Transverse waves

In a water wave the particles are moving up and down, while the wave is moving along. In such a case as this where the displacement is at right angles to the movement, the wave motion is called transverse.

(b) Longitudinal waves

If the engine of a shunting train jerks, the trucks show a wave motion in which the vibration and the wave direction are in the same line. This is called a longitudinal wave motion.

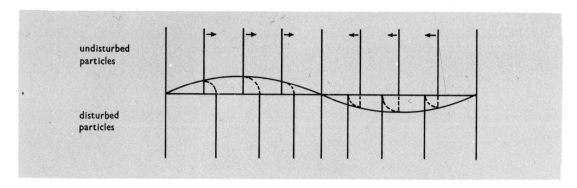

Fig. 161   The representation of a longitudinal wave by a sine curve

The connection with a sine curve is shown above. Either of these types of waves may exist as progressive or stationary waves.

(c)   Progressive waves

When a wave-motion begins to be propagated the edge of the disturbance travels outwards. It is called a progressive wave, the outline and energy of which is continuously advancing. Neighbouring points on this type of wave will be out of step (or out of phase) with each other.

(d)   Stationary waves

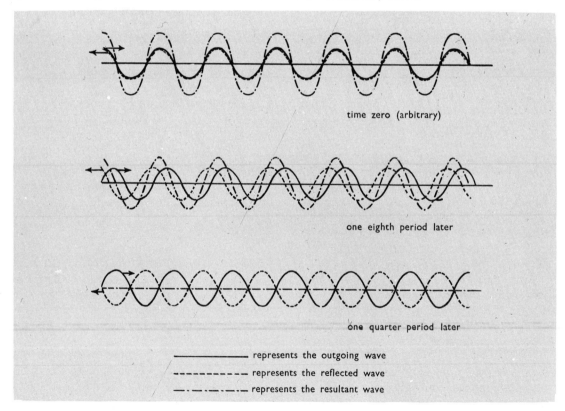

time zero (arbitrary)

one eighth period later

one quarter period later

——————— represents the outgoing wave

------------- represents the reflected wave

—·—·—·— represents the resultant wave

Fig. 162   Stationary wave

When a progressive wave reaches the edge of the medium in which it is travelling it will be (partially) reflected. As the outgoing and reflected waves cross they set up a distinctive pattern called a stationary wave.

The resultant is found by adding the other two waves together.

Note.

(*i*) There are a number of equally spaced points (nodes) where the medium is always undisturbed.

(*ii*) Between the nodes the particles have different amplitudes of vibration varying from zero at the node to twice the amplitude of the constituent waves at the antinode. All the particles in the space between two nodes move in step (phase) so that they all reach their respective amplitudes at the same instant.

(*iii*) The wavelength of the stationary (standing) wave is the same as that of the other two waves.

30.4   Stationary waves on a string. Melde's experiment

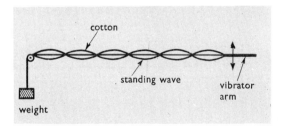

Fig. 163   Melde's experiment

A light cotton is attached to the arm of a vibrator (driven either from a signal generator or low voltage a.c. derived from the mains supply) passed over a pulley and kept taut by a weight as shown in the diagram. The cotton will show a stationary transverse wave motion if a suitable frequency (or load) is chosen.

It can be shown that the velocity (*v*) of transverse waves is related to the tension (*T*) and the mass per unit length (*m*) of the cotton, by the equation

$$v = \sqrt{\frac{T}{m}}$$

so if the length of the cotton from pulley to vibrator is *l* and *n* loops are formed, the wavelength (λ) of the stationary wave is $2l/n$ so the frequency *f* is given by

$$f = \frac{v}{\lambda} = \frac{n}{2l}\sqrt{\frac{T}{m}}$$

so the value of *f* may be readily determined

30.5   The sonometer

The lowest frequency of transverse vibration of a stretched string is, from the above,

$$f = \frac{1}{2l}\sqrt{\frac{T}{m}}$$

This equation may be verified with the sonometer.

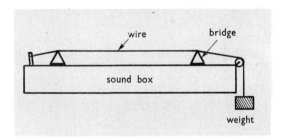

Fig. 164   The sonometer

A wire is stretched across two movable bridges and kept taut by the weight. It is tuned to be in unison with various tuning forks, either by observing beats between the plucked wire and the sounding fork, or by placing a small paper rider on the wire and adjusting until the rider is thrown off by resonance when a sounding fork is placed on the sound box of the sonometer.

(i)   To verify $f \propto 1/l$ at a fixed T and m. The resonating length is found for several forks. A graph of f against $1/l$ should be a straight line.

(ii)   To verify $f^2 \propto T$ at a fixed m and l. The tension is found for a match with various forks. A graph of f^2 against T should be a straight line.

(iii)   To investigate the effect of mass per unit length various wires of the same material but of different diameters are used. As neither the fork frequency nor the mass per unit length can be continuously varied, the tension is kept constant but the length has to be found for each wire to be in resonance with the same fork. Using the result (i) the frequency is calculated for a fixed length of wire and hence a graph of the calculated frequency squared against $1/m$ should be a straight line.

31   THE REFLECTION AND REFRACTION OF WAVES

31.1   The reflection of waves

The reflection of waves can be shown for each type as follows:

(i)   Water waves—in a ripple tank with a barrier

(ii)   Sound waves—

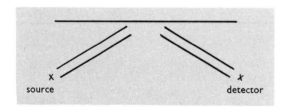

Fig. 165   Reflection of sound waves

(iii) Radio waves—with a 3 cm emitter, metal reflector and a suitable detector.

(iv)   Light waves—with a mirror and a ray box.

In each case it is found that the angle of incidence is equal to the angle of reflection.

The general theory of wave reflection is based on Huygen's idea of secondary wavelets.

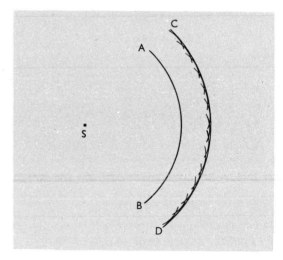

Fig. 166   Huygen's secondary wavelets

Suppose that AB is, at some instant, the front of the disturbance originating from S. Then every point on AB acts as a source of wavelets, drawn as shown. The envelope of these, CD, is the new wavefront.

Note that if AB is a long way from S it becomes indistinguishable from a straight line, (or a plane in three dimensions) and is called a plane wavefront.

N.B. The ray used in geometrical optics is perpendicular to the wavefront.

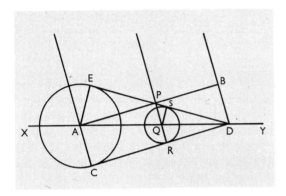

Fig. 167 Wave theory of reflection

Imagine a plane wave AB reaching a plane reflecting boundary XY. In the time it takes for wavelets from B to reach D, those from A have spread over a circle radius AC. CD is the position that the wavefront would reach if XY were not there.

Likewise at P the wavelets travel normally to Q, then spread out over a circle radius QR.

The reflected wave is the envelope ESD.

As $QS = QR$ radii $QD = QD$, and $Q\hat{S}D = Q\hat{R}D$ rt. angles $\triangle$'s QSD and QRD are congruent,

$\therefore S\hat{D}Q = Q\hat{D}R$ but $Q\hat{D}R$ is fixed $\therefore S\hat{D}Q$ is fixed for all positions of S (and therefore P) so ESD must be a straight line.

Therefore a plane wave reflects from a plane surface as a plane wave.

By geometry

$$B\hat{A}D = E\hat{D}A \begin{cases} \text{as } B\hat{A}D = A\hat{D}C \text{ alternate} \\ \qquad = E\hat{D}A \text{ proved} \end{cases}$$

and as $B\hat{A}D$ = angle of incidence
and $E\hat{D}A$ = angle of reflection

(N.B. the ray is perpendicular to the wave-front) these angles are proved equal.

31.2   The refraction of waves

The refraction of waves can be shown for each type as follows:

(*i*)   Water waves – in a ripple tank with a shallow region.

(*ii*)   Sound waves – a balloon filled with carbon dioxide will act as lens.

(*iii*)   Radio waves – 3 cm waves can be refracted through blocks of paraffin wax.

(*iv*)   Light waves – a perspex block and a ray box.

The general theory of refraction can again be derived from Huygen's secondary wavelets.

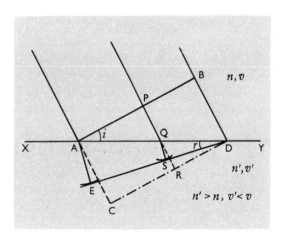

Fig. 168   Wave theory of refraction

Imagine a plane wave AB reaching a plane refracting surface XY.

As the waves travel more slowly in the denser medium they reach only a circle radius AE in the time t it takes for wavelets to go from B to D.

and $$\frac{AE}{AC} = \frac{v't}{vt} = \frac{v'}{v}$$

Likewise for intermediate positions,

$$\frac{QS}{QR} = \frac{v'}{v}$$

so $$\frac{QS}{QR} = \frac{AE}{AC}$$

$$\therefore \quad \frac{QS}{AE} = \frac{QR}{AC} = \frac{QD}{AD},$$

by similar $\triangle$'s ACD, QRD.

$\therefore \quad \triangle$'s AED and QSD must be similar

so ED must be a straight line, which is the envelope of the refracted wave. So a plane wave refracts as a plane wave.

Now $\overset{\wedge}{BAD}$ = angle of incidence, i

$\overset{\wedge}{ADE}$ = angle of refraction, r

and $\sin i = \dfrac{BD}{AD} = \dfrac{AC}{AD}$

$$\sin r = \frac{AE}{AD}$$

$$\therefore \quad \frac{\sin i}{\sin r} = \frac{AC}{AD} \bigg/ \frac{AE}{AD} = \frac{AC}{AE} = \frac{v}{v'} = \frac{n'}{n}$$

a constant.

If the first medium is a vacuum.

$$v = c$$
$$n = 1$$

$$\therefore \quad \frac{c}{v'} = n'$$

or $$v' = \frac{c}{n'}$$

thus the speed of light in a given medium can be found. Also as the frequency is unchanged (shown by the colour being constant) the wavelength is also less in a denser medium.

Note that if $v' > v$, AE can be greater than AD and no envelope can be drawn in the second medium, so total internal reflection occurs.

32   THE SUPERPOSITION OF WAVES

When two similar waves cross, the resultant displacement can always be found by adding the separate displacements of the two component waves together as in the case of a stationary wave.

Various cases can be distinguished.

32.1   Equal frequency, equal amplitude

Over a certain region it is possible for waves from two sources to be in step (phase) in some places and out of step in others. At the former the amplitude of vibration will be twice that of either wave, at the latter places there will be zero vibration.

Thus if two waves of this type cross, there will be a redistribution of energy. This is called an interference pattern, and the phenomenon itself is called interference.

For the interference pattern to be stable,

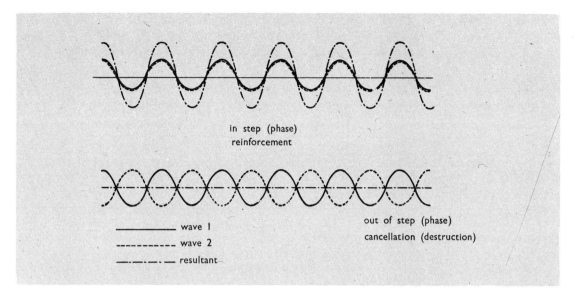

in step (phase)
reinforcement

out of step (phase)
cancellation (destruction)

——————— wave 1
------------- wave 2
——·—·— resultant

Fig. 169 Superposition of waves of equal frequency and amplitude

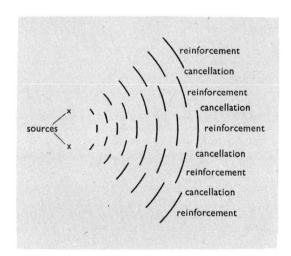

reinforcement
cancellation
reinforcement
cancellation
reinforcement
cancellation
reinforcement
cancellation
reinforcement

sources

Fig. 170 Pattern resulting from the superposition of waves of equal frequency and amplitude

and therefore detectable, the two sources of the waves must be coherent, i.e. in a permanent phase relation to one another. If the waves are pulsed—as in the case of light—this cannot be achieved with two separate sources so the light from one source has to be split into two coherent parts.

From the diagrams note that:

Reinforcement occurs for the waves $n\lambda$ out of step.

Destruction occurs for the waves $(n \pm \frac{1}{2})\lambda$ out of step where n is an integer.

Interference can be observed with all the types of waves under consideration.

(i) Water waves – In a ripple tank with
(a) Two point dippers.
(b) A plane wave and a double slit.

(ii) Sound waves – Quincke's tube.

(iii) Radio waves – 3-cm waves show
(a) Double slit type.
(b) Lloyd's mirror type.

(iv) Light waves – (a) Young's double slit.
(b) Fresnel biprism.
(c) Lloyd's mirror.
(d) Thin films.
(e) Newton's rings.

Comments on the various methods

(*ia*) The two dippers vibrate together, therefore the vibrations produced by them must be coherent, and if BP − AP = odd

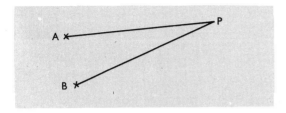

Fig. 171 Interference from two point dippers in a ripple tank

number of half-wavelengths there is no disturbance and if BP − AP = even number of half-wavelengths there is maximum disturbance, if A and B are in phase with each other.

(*ib*), (*iiia*), (*iva*), (*ivb*) The double-slit type of interference must, in the case of light, be studied quantitatively, so the theory will be worked out.

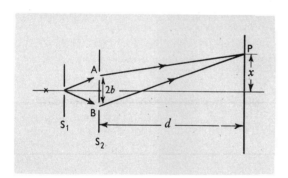

Fig. 172 Double slit type of interference

A point or line source is placed near to a single narrow slit S_1 followed by S_2, (two parallel slits about 1 mm apart.) Fringes are observed on a screen, or through an eyepiece at P.

If white light is used various coloured fringes are observed. The pattern has a white centre on the axis corresponding to no path difference.

If monochromatic light is used, then dark and bright line fringes are observed. These are more suitable for measurement.

$$\left. \begin{array}{l} AP^2 = d^2 + (x-b)^2 \\ BP^2 = d^2 + (x+b)^2 \end{array} \right\} \text{ by Pythagoras}$$

$$\therefore \quad BP^2 - AP^2 = 4bx$$

but $$BP + AP \simeq 2d$$

$$\therefore \quad BP - AP = \frac{4bx}{2d} = \frac{2bx}{d}$$

If P is the nth dark fringe from the centre,

$$\frac{2bx_n}{d} = (n - \tfrac{1}{2})\lambda$$

$$\therefore \quad x_n = \frac{d(n - \tfrac{1}{2})\lambda}{2b}$$

and for the $(n + m)^{\text{th}}$ fringe

$$x_{n+m} = \frac{d(n + m - \tfrac{1}{2})\lambda}{2b}$$

$$\therefore \quad x_{n+m} - x_n = \frac{dm\lambda}{2b}$$

so fringes are evenly spaced, as the separation is independent of n.

N.B. Fresnel's biprism is an alternative way of obtaining the two coherent sources.

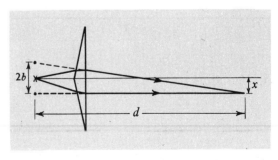

Fig. 173 Fresnel's biprism

The same expression holds for the light coming from the two virtual images formed by the biprism and setting up the interference pattern.

(*ii*)

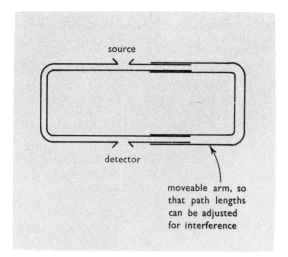

Fig. 174   Quincke's tube

(*iiib*), (*ivc*)

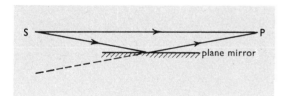

Fig. 175   Lloyd's mirror

Fringes at P are due to interference between waves travelling direct from S and those reflected at grazing incidence.

With white light the centre fringe is black so all waves must be out of step. This is due to a phase change of π which always occurs on reflection at a denser medium, which, in this case, is the mirror.

(*ivd*)   Thin films, such as oil on water, or soap bubbles, are often coloured due to interference between light reflected from the top and bottom of the layer. These colours arise because certain other colours (wavelengths) of the original white light are suppressed by interference.

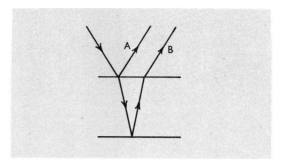

Fig. 176   Thin film interference

There is a path difference between the rays A and B and if these rays are made to overlap interference results.

(*ive*)   This is a special case of (*ivd*), where the film is a curved air film produced between a long-focus convex lens and a plane surface. By symmetry, rings must be formed about the point of contact.

32.2   Near equal amplitude, unequal frequency

This case is usually important when the two waves are only slightly different in frequency.

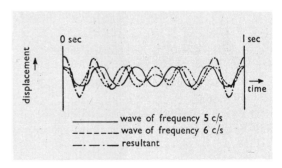

Fig. 177   The formation of beats

The resultant shows a varying amplitude for any particle with time. The variation is a series of maxima and minima, which are called beats. They have a frequency equal to the difference in frequency of the two original waves.

Beating can be shown by

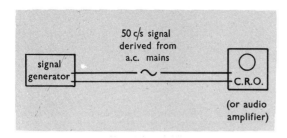

Fig. 178 Beats from electrical signals

(*i*) Using two similar tuning forks, one lightly loaded with plasticine.

(*ii*) Combining two a.c. electrical signals.

32.3 Unequal amplitude, multiple frequency

For example, suppose a main wave and one of twice the frequency but smaller amplitude are added together.

The resultant has the same frequency but a

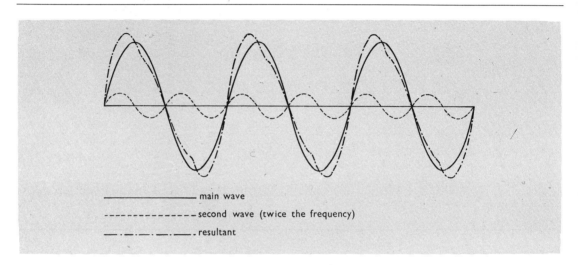

—————————— main wave

— — — — — — — second wave (twice the frequency)

—·—·—·—·—·— resultant

Fig. 179 The effect of harmonics on wave outline

different shape from the main wave. Further multiple frequencies can be added in with similar effects.

This is the reason for the different sounds of musical instruments when playing the same note. All sound the same fundamental frequency (first harmonic) but the higher multiple frequencies (second, etc. harmonics) are present to different extents, altering the resultant wave shape and 'quality' of the note.

33 THE VELOCITY OF WAVES

The measurement of this quantity is of great importance.

33.1 The velocity of sound in a tube

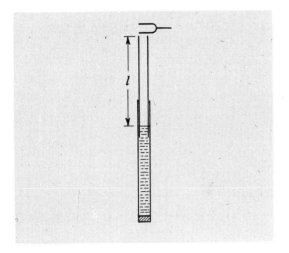

Fig. 180　A resonance tube

Two tubes, one filled with water and the other moving inside it, are adjusted for resonance with a tuning fork held as shown. The length l is kept as short as possible.

It is found that the length l is not the true length of the resonating column so that an end correction e is added to l.

A second resonant position at about three times the length can also be detected for the same fork.

These stationary waves are shown diagrammatically, with the nodes and antinodes being placed as shown.

(N.B. The air is not moving from side to side but longitudinally).

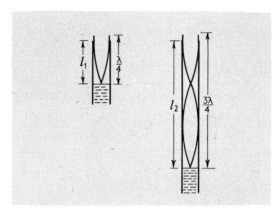

Fig. 181　Stationary waves in an air column

The results can be treated in two ways:

(*i*)　With one fork and two resonance positions.

$$l_1 + e = \frac{\lambda}{4} \quad l_2 + e = \frac{3\lambda}{4}$$

$$\therefore \quad \frac{\lambda}{2} = l_2 - l_1$$

If $v =$ velocity of sound and $f =$ frequency of the tuning fork, then $v = f\lambda = 2f(l_2 - l_1)$.

(*ii*)　With several forks and the first resonance position for each.

$$\frac{\lambda}{4} = l + e \quad \text{but} \quad v = f\lambda$$

$$\therefore \quad \frac{v}{4f} = l + e$$

$$\therefore \quad l = \frac{v}{4}\frac{1}{f} - e$$

So if l is plotted against $1/f$ the gradient is $v/4$.

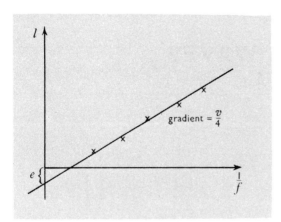

Fig. 182 Graphical determination of velocity of sound from resonance tube measurements

N.B. e is approximately $0.58 \times$ radius of tube.

33.2 The velocity of sound in open air. (Hebb's method)

Two large paraboloidal mirrors were used. A microphone was placed at the focus of each and connected to a primary of a two-primary transformer. A whistle was blown at one focus and the resultant sound was heard on headphones. One reflector and its microphone are moved and the combined signals show maxima and minima. The distance between successive minima gives the wavelength of the sound, and from the known frequency the velocity may be found. Hebb worked through about 200 minima (about 100 ft). The whistle was compared to a standard tuning fork.

The particular advantages are:

(*i*) It is a relatively compact apparatus. The experiment was originally performed in a hall so that the temperature and humidity could be observed.

(*ii*) It is a free air method, so no 'tube corrections' are needed.

(*iii*) There are no errors due to a large intensity of sound.

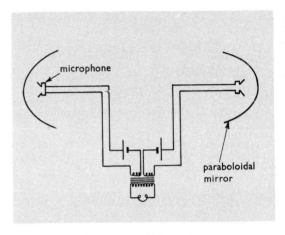

Fig. 183 Hebb's method

33.3 The velocity of sound in a rod (Kundt's tube)

A long glass tube is closed at one end by a movable piston and at the other end by a rod clamped at its mid-point. The tube, which must be carefully dried, contains a little lycopodium powder or cork dust. When the wooden rod is set into longitudinal vibration

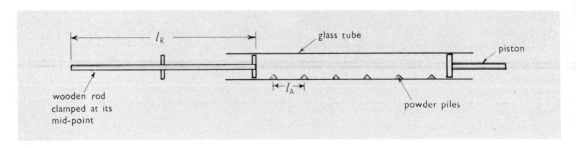

Fig. 184 Kundt's tube

by pulling along it with a resined cloth and the piston adjusted so that a stationary wave is set up in the air in the tube, the powder gradually gathers into small heaps at the nodes. Thus the internodal distance l_A can be found.

The wavelength of the sound in the rod is equal to twice the length (l_R) of the rod. If the velocity of sound in the rod is v_R and in air is v_A, then the frequency of the note is given by

$$\frac{v_R}{2l_R} \quad \text{and by} \quad \frac{v_A}{2l_A}$$

so

$$v_R = v_A \cdot \frac{l_R}{l_A}$$

The experiment may be extended to find Young's modulus (E) for the wood (using the formula $v_R = \sqrt{E/\rho}$ where $\rho =$ density of the wood), or to find the velocity of sound in other gases if the tube is filled with them.

33.4   The velocity of light (c)

The velocity of light is a very important constant in modern theory, and probably the most accurate method of determining it was due to Michelson.

Light from the source O is reflected from a highly polished octagonal steel prism which can be rotated at high speed. The light then

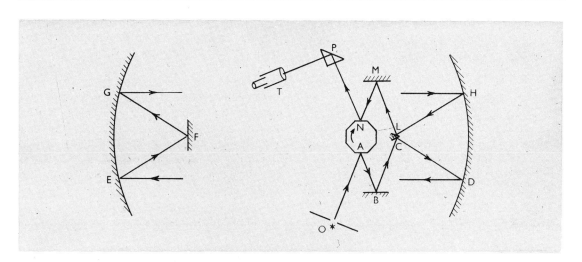

Fig. 185   Michelson's method

follows the path BCDEFGHLMN back to the prism. L, C and F are at the principal foci of their respective spherical mirrors which are a long way apart (≈ 22 miles). From the steel prism the light enters a telescope T by way of a totally reflecting prism P. When the steel prism is rotated the image is moved unless the prism makes $\frac{1}{8}$th of a revolution in the time the light takes to travel from A to N.

Suppose the octagonal prism makes n rev/sec, the total distance A to N is d,

then

$$c = \frac{d}{1/8n} = 8nd$$

The great virtue of the method is that it is a null method.

(1926 result $- 299{,}795 \pm 4$ km/s)

33.5   The variation of sound wave velocity

The velocity in a gas is proportional to the square root of the absolute temperature, but independent of the pressure. If a wind is blowing the velocity of the sound wave is altered by adding the component of the wind in the direction of the sound wave.

34   THE FREQUENCY OF SOUND WAVES. THE DOPPLER EFFECT

34.1   The frequency of sound waves

In order to find the frequency of a source of sound it is usually compared with a source of known frequency in one of several ways.

(*a*)   *C.R.O. and signal generator*
A double-beam oscilloscope is needed.

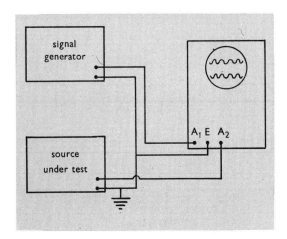

Fig. 186   Comparison of frequencies on a double beam C.R.O.

Adjust the signal generator until there is the same spacing on each trace.

(*b*)   *Stroboscope*
Illuminate the vibrating object strobo-scopically and increase the strobe' frequency for the last stationary position. The two frequencies are then equal.

(*c*)   *Beats*

If the frequency of the unknown source and the calibrated source are close, beats will be heard. As the difference between the frequencies equals the beat frequency, the unknown frequency can be found. In order to decide which of the two possible values (i.e. known frequency $\pm$ beat frequency) is correct the known frequency is changed slightly and the effect on the beat frequency is observed.

34.2   Some available sources of known frequency

(*i*)   Tuning fork – The maker's calibration can usually be accepted.

(*ii*)   Audio signal generator – This can be calibrated against the mains frequency by Lissajous' figures.

(*iii*)   Siren – The frequency is found mechanically. (The number of holes multiplied by the number of rev s^{-1}).

34.3   The Doppler effect

Whenever a source of waves or the observer

are in motion there is a change in frequency compared to that observed when there is no movement. If they approach, the frequency is increased, and if they are moving apart, it is decreased.

(a) *An observer approaching a stationary source*

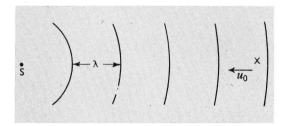

Fig. 187 Doppler effect, moving observer

Let the wave velocity be v, the emitted frequency f, and the emitted wavelength λ.

Then a stationary observer receives v/λ waves $s^{-1} = f$, but if he moves a distance u_0 in one second, (i.e. his velocity equals u_0) then he receives an extra u_0/λ waves s^{-1},

so new frequency

$$f' = \frac{v}{\lambda} + \frac{u_0}{\lambda} = \frac{v + u_0}{\lambda} = \frac{v + u_0}{v/f}$$

$$\therefore \quad f' = \frac{v + u_0}{v} \cdot f$$

If the observer recedes, then u_0 takes a minus sign.

(b) *Source approaching stationary observer*

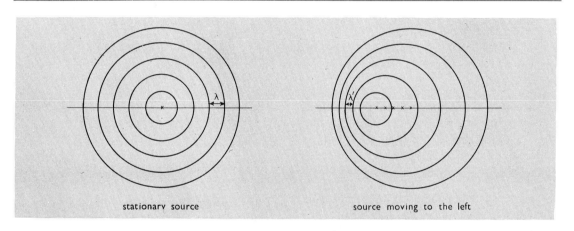

Fig. 188 Doppler effect, moving source

By comparing the wavefronts shown in the diagram the approaching source gives rise to a shorter wavelength.

For a stationary source in 1 s f waves are emitted covering a distance v, but for the moving source these same waves only cover a distance $v - u_s$ where u_s = the velocity of the source.

$$\therefore \quad \frac{\lambda'}{\lambda} = \frac{v - u_s}{v}$$

but $\quad v = f\lambda = f'\lambda'$
where f' = the observed frequency

$$\therefore \quad \frac{f}{f'} = \frac{v - u_s}{v}$$

$$f' = \frac{v}{v - u_s} \cdot f$$

Note that if the source and the observer are both moving towards each other

H

$$f' = \frac{v + u_0}{v - u_s} \cdot f$$

In the case of light, the theory of relativity must be used to give a correct interpretation, and

$$f' = \frac{1 + (u/c)}{\sqrt{1 - (u^2/c^2)}} \cdot f$$

where u = the relative velocity of approach
c = the velocity of light

and the effect shows up in the red shift, i.e. the spectrum from stars is shifted to the red end of the spectrum (longer wavelength, lower frequency). Therefore u is negative and the recessional velocities can be calculated.

35 DIFFRACTION AND POLARIZATION

35.1 Diffraction

Under certain conditions waves do not travel with straight boundaries, but spread out into other regions. This is called diffraction and can occur with all types of waves.

(a) Water waves

A plane wave is generated in a ripple tank so that it is incident on an adjustable slit.

If the slit is originally wide compared to the wavelength then little diffraction occurs. As the slit is narrowed, diffraction increases. If the wavelength is then shortened diffraction decreases again showing that for maximum effect the size of the obstacle and the wavelength should be about equal.

(b) Sound waves

Diffraction occurs very readily with ordinary objects, as the wavelength is about one metre for frequencies of about 300 Hz. The most obvious example is that people can hear round corners.

(c) Radio waves

Using a 3 cm radio wave emitter and detector the first part of (a) can be repeated.

(d) Light

Diffraction fringes can be observed when light passes

(i) through a single slit,
(ii) by a straight edge,
(iii) through a small hole when the right diameter can cause almost darkness at a chosen point on the axis of the hole,
(iv) by a small circular object which will give a bright spot on the axis.

35.2 The diffraction grating

Probably the most important case of the diffraction of light occurs when parallel rays fall on to a grating, a series of very close parallel slits. These are produced by ruling lines on glass, but cheaper plastic replicas are commonly used.

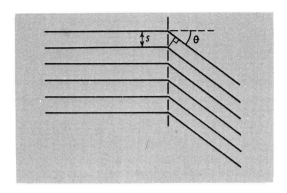

Fig. 189   Theory of the diffraction grating

When the light reaches the grating, diffraction causes a spreading out, but usually random interference means that nothing is observed.

However, in some directions all the waves can be in step.

This occurs when

$$s \sin \theta = n\lambda \quad n = \text{integer}$$

The integer n is called the 'order' of the diffraction image. Only a finite number of directions in which the waves are in step are possible in a particular case as $\sin \theta$ cannot exceed 1.

Measurements can be made with a diffraction grating on a spectrometer, and the wavelength of the incident light can be found.

The preliminary adjustments of the spectrometer must be repeated as described earlier. Then the grating must be placed exactly perpendicular to the light coming from the collimator. To achieve this the telescope is placed in the straight through position and then rotated through exactly 90° using the verniers. The grating is then placed on the table and adjusted so that a reflected image of the slit appears to coincide with the crosswires of the telescope. The table is then turned through exactly 135°, using the verniers again, to leave the grating in the required position with the glass backing towards the collimator.

The telescope can now be used to view the various order spectra and the required angles may be readily measured.

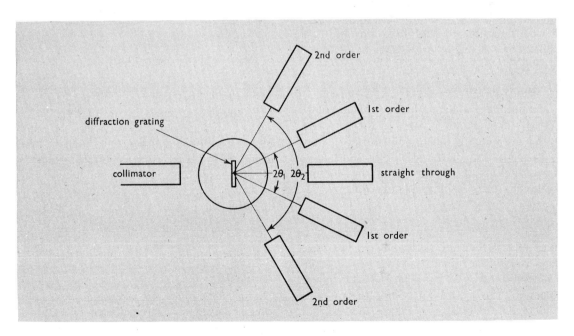

Fig. 190   The spectrometer used with a diffraction grating

35.3 Polarization

A transverse wave usually has components vibrating in every direction in a plane perpendicular to the direction of travel, but under certain circumstances the vibrations are restricted to one direction and the wave is said to be plane polarized.

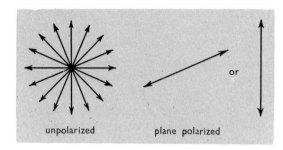

Fig. 191 Polarization

Polarization can occur in various cases.

(a) Waves on a stretched 'string'

A length of rubber tubing is stretched through two slits (made from half-metre rules), and the end 'rotated'. Vibrations are restricted to one direction by the first slit and are stopped by the second if it is at right angles to the first.

(b) Radio waves

The output of a 3-cm emitter is plane polarized. If the detector is rotated the received signal varies greatly due to polarization.

Especially in short-wave radio and television the aerial has to be in the correct position to receive polarized radiation.

(c) Light

Light which has been passed through a sheet of polaroid and then falls on a second sheet shows the same behaviour as the other waves above. Light can therefore be 'plane polarized'. (Polaroid is an assembly of quinine iodosulphate crystals, all in the same direction, which splits light into two polarized parts which are polarized at right angles to each other absorbing one and transmitting the other).

Light reflected from a glass plate at an angle of incidence of about 56° (where $\tan \theta = n$) is also polarized and the transmitted ray is also partially polarized. A pile of plates gives better polarization.

When objects are viewed through certain crystals (e.g. calcite) two images can be seen. There are two refracted rays originating from one incident ray. These refracted rays are polarized in directions which are at right angles to each other.

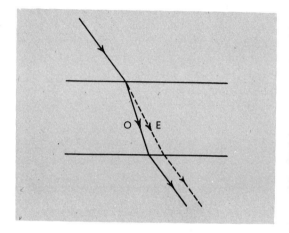

Fig. 192 Double refraction

O is the ordinary ray $n = 1.658$
E is the extraordinary ray $n = 1.486$
E is so called because with the right setting E will rotate about O as the crystal is turned. This is referred to as double refraction.

Note that two devices are needed to experiment with polarization, one to produce it and the other to detect it. The first is called a polarizer and the other an analyser.

35.4 Some applications of polarized light

(a) Polarimetry

Some substances, for example sugar solutions, are optically active, i.e. rotate the plane of polarized light. If they are placed between crossed polaroids, the rotation necessary to return to extinction is found, and as this depends on the strength (etc.) of the solution useful information can be found.

(b) Photoelasticity

Some materials (for example certain plastics) are doubly refracting under strain, and set up coloured fringes when viewed between crossed polaroids. Models can thus be tested for engineering purposes.

ELECTRICITY AND MAGNETISM

36 MAGNETISM

36.1 The elementary facts

(*i*) A magnet attracts small pieces of iron and steel, which stick to it, especially in certain regions, the poles of the magnet.

(*ii*) If a magnet is freely suspended it comes to rest pointing approximately North–South, hence the poles are named North and South poles.

(*iii*) Other magnets are attracted or repelled, but only repulsion is a sure test for magnetization. Like poles repel, unlike poles attract.

(*iv*) The effects of the magnet extend over a region round the magnet called the magnetic field. It can be mapped by iron filings or a plotting compass and represented by lines of force.

(*v*) Pieces of soft iron placed near a magnet become temporary magnets (by induction).

36.2 The production of magnets

(*i*) By single/double touch, but these methods can only produce weak magnets.

(*ii*) With a solenoid using direct current, when only a momentary large current is needed.

(*iii*) With a solenoid using alternating current. This is a process based on chance, the operator hoping to switch off when the current is near its maximum value.

36.3 Demagnetization

(*i*) Banging the magnet when it lies East–West.

(*ii*) Heating the magnet.

(*iii*) Placing it in a decreasing, alternating magnetic field (i.e. in a solenoid with a decreasing alternating current, or moving it out along the axis when a constant alternating current is flowing.

36.4 Terrestrial magnetism

If a magnetized needle is freely suspended by a torsionless thread through its centre of gravity, it will come to rest in a definite direction, and at a definite angle to the horizontal.

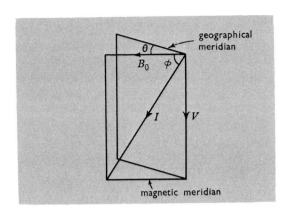

Fig. 193   The magnetic elements

where θ = declination (variation)
ϕ = dip
B_0 = horizontal component ⎫ of earth's
V = vertical component ⎬ magnetic
⎭ field
I = total field

θ is measured directly as the angle between the meridians,

ϕ is measured by the dip circle,

B_0 can be measured with a deflection magnetometer and another known field, or with a ballistic galvanometer.

The magnetic elements vary from place to place and with time.

37 QUALITATIVE ELECTROSTATICS

37.1 Electrification by friction

Many objects when rubbed have the power of attracting light pieces of cork, paper or pith. However, once the objects have touched the rubbed material they are repelled from it.

If a rod of polythene is rubbed with a duster and placed on a pivoted support then it will move away from a similar rubbed rod brought near to it, but will be attracted by a perspex rod rubbed with a duster.

These facts can be explained by stating that the bars become electrified, that is, they acquire electric charge. A charged object attracts neutral ones, until they touch it. Then they acquire some of its charge and are repelled.

From the second experiment it can be concluded that two states of electrification exist. It has been agreed to call the charge on polythene, when it has been rubbed, negative; and that on the perspex, when it has been rubbed, positive.

Thus like charges repel, unlike charges attract.

37.2 Insulators and conductors

Metal rods held in the hand and rubbed, show no signs of electrification. This is because any charge they acquire can flow through them and the human body to the earth. Hence they are called conductors. Materials like ebonite and silica, where the charge cannot move in this way are called insulators. For insulators to be really effective they and the air around them must be dry.

A piece of conducting material mounted on an insulating handle can acquire a charge as the charge cannot flow away through the insulator.

37.3 The mechanism of the production of charge by friction

If a polythene rod is charged by rotating a woollen cap held by a silk thread at one end, then the rod will show a negative charge, the cap a positive one, and the two together no resultant charge. This shows that positive and negative charges are formed in equal amounts by friction.

This can be explained in a simple way if some negative electrons move from the wool to the polythene, leaving the wool with excess positive nuclei, and therefore with a positive charge.

The transfer occurs when the two materials are in contact, rubbing only increases the effective area of contact.

37.4 Charging by induction

A charged rod is brought near to an insulated conductor, and a charge distribution is set up as shown at (i).

The conductor is then momentarily earthed, the positive charge is removed, but the negative charge is still 'bound' (ii).

On removing the charged rod the negative charge can spread out over the whole conductor (iii), which is then said to be charged by induction.

Note that no charge leaves the rod, but energy is not created, as work has to be done in order to separate the charged rod and the charged body.

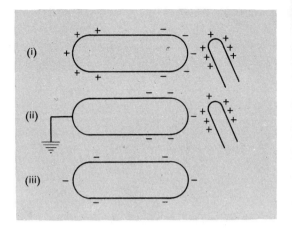

Fig. 194   Charging by induction

37.5 The electrophorus

The electrophorus consists of an ebonite plate mounted on a metal sole, and a brass disc, mounted on an insulating handle, which can be placed on the ebonite. When one is placed on the other, contact is made only at a few points, as indicated in the diagram.

The ebonite is charged by flicking it with fur as shown at (i). When the disc is placed on the ebonite charges are induced on it as shown at (ii). Charging is by induction and not by conduction as the area of contact is small.

Momentarily earthing the brass removes the negative charge from it (iii). When the two are separated the brass is positively charged (iv) and the negative charge on the

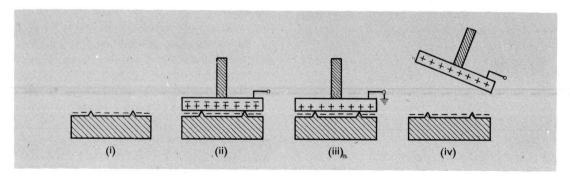

Fig. 195   The electrophorus

ebonite has not been reduced. Thus the process can (theoretically) be repeated indefinitely.

37.6   Potential (qualitative)

Just as temperature, water level or air pressure, govern the direction of flow of heat, water or air, electrical potential governs the direction of flow of electricity. The earth is always taken to be at zero potential, and objects will be at positive or negative potential if (positive) electricity flows from or to them respectively when they are metallically connected to earth.

The case of charging by induction can be explained by these ideas.

The charged rod is at a positive potential due to its own charge. This makes the end of the conductor nearer the rod at a higher positive potential than the far end. Electricity flows through the conductor to give equal potential all over the surface, resulting in the charge distribution shown. When the conductor is earthed the potential must fall to zero, and the negative charge neutralizes the positive potential due to the positive rod.

37.7   The gold leaf electroscope

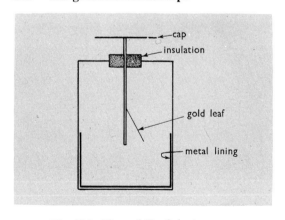

Fig. 196   The gold leaf electroscope

This consists of a box with a glass front and back. If the box is not made of metal it will have a strip or lining of metal. A metal plate (the 'cap' of the electroscope) is mounted on a metal rod, which is carefully insulated from the box. The bottom end of the rod is flattened and a piece of gold leaf is fastened to it.

The instrument is one which can be used primarily to detect potential. The following experiments illustrate its use.

(*i*) The cap is touched with a charged object. The leaf is deflected, because it has been charged, and so there will be a difference in potential between it and the case.

(*ii*) The electroscope is insulated, the cap earthed, and a charged object touched on to a terminal connected to the lining, again the leaf is deflected owing to a difference in potential.

(*iii*) A charged object is brought near to the cap. The leaf is deflected while the charged object is near as a potential difference is set up, although no charge is permanently transferred to the cap.

(*iv*) The cap and lining are joined by a wire. If a charged object is allowed to touch the cap there is no deflection, as there is no potential difference, although charge is transferred.

A gold leaf electroscope can be used to test for charge, and the sign of charge. It must first be given a charge of known sign, for example, positive. When a positively charged object is brought near, the leaf is deflected more, as a greater potential difference will exist between leaf and case. When a negatively charged object is brought up, the potential difference, and therefore the deflection is reduced, (although the charge is unaltered). If the object is brought close enough, the leaf will collapse entirely, and then may diverge again, (the potential passes through zero to negative). Uncharged bodies will also cause the deflection to decrease so a decreased deflection is not a satisfactory test for charge.

The gold leaf electroscope can be charged by contact (conduction) or induction. With a positively charged object, the first method gives a positively charged electroscope, the second negatively charged.

37.8 The position of charge on a conductor

It can be shown that the charge on a conductor resides entirely on the outside surface.

(a) Biot's experiment

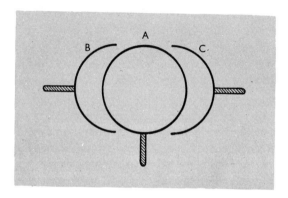

Fig. 197 Biot's apparatus

A is a charged insulated brass sphere, B and C two hemispherical metal caps fitting on to A, each fitted with insulating handles.

If B and C are fitted on to A, then removed and tested for charge, A will be uncharged while B and C are charged.

If the three parts are uncharged, but together, and then a charge is given, and the three separated, it will be found that the charge is entirely on B and C.

(b) Faraday's butterfly net experiment

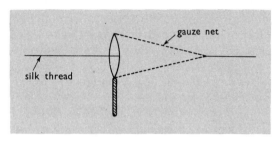

Fig. 198 Faraday's butterfly net

A conical net of linen gauze is mounted on an insulating stand, so that it can be turned inside out by the insulating silk threads. The net is charged and tested, and charge is only found on the outside; the net is pulled inside out, and the charge is again found to be on the (new) outside.

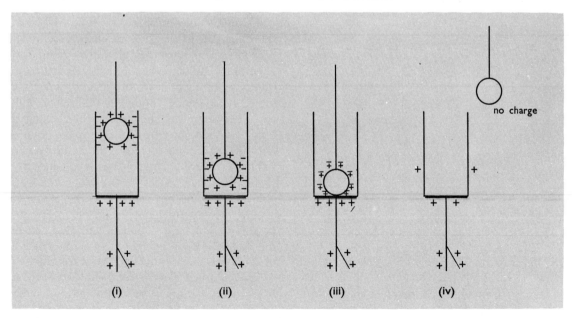

Fig. 199 Faraday's ice pail experiment

37.9   Faraday's ice pail experiment

A charged sphere is lowered into a tall can standing upon a gold leaf electroscope. Induced charges appear as shown, resulting in a deflection of the leaf (i). Once the sphere is well inside the can there is no change in the deflection however the sphere is moved about (ii).

Also if the sphere touches the can (iii) there is no change in deflection, and subsequently the sphere is found to be uncharged (iv).

This is explained if the induced charges are equal to one another, but of opposite sign, and also equal to the inducing charge.

Thus the experiment shows the relationship between the inducing charge and the total induced charge.

The experiment also shows that all the charge can be collected from an object by touching it to the inside of a conductor, large enough to embrace all the charge.

37.10   The distribution of potential and charge over the surface of a conductor

(a)   Potential

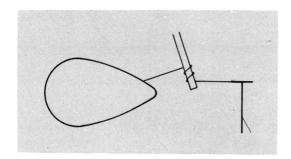

Fig. 200   Distribution of potential

A pear-shaped conductor is charged, and then joined by a wire to a gold leaf electroscope. The wire is wound round an insulating handle so that it can be moved into contact with any part of the conductor. A constant deflection is obtained, indicating a constant potential all over the surface of the conductor.

(b)   Charge

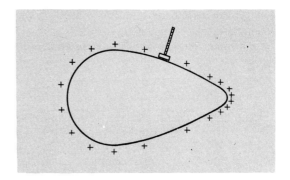

Fig. 201   Distribution of charge

For this a proof plane is used. This is a small brass disc fitted to an insulating handle. It is brought into contact with part of the conductor and acquires a charge by conduction proportional to the surface density of charge at that part of the conductor. It is then touched to the inside of a can resting on a gold leaf electroscope, giving up all its acquired charge. The deflection of the leaf is therefore a measure of the charge. The experiment can be repeated, using different parts of the conductor. It is found that the regions where the radius of curvature of the surface is smallest have the greatest charge density.

37.11   The discharging action of points

At corners and points the charge density is very great and at such regions on the

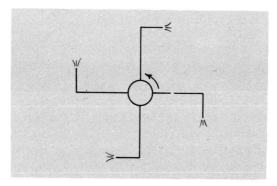

Fig. 202   The electric windmill

surface considerable charge leakage occurs. This is illustrated by the electric windmill which turns under the reaction caused by the discharge.

Lightning conductors also apply this principle, charge leaking to reduce the charge on the thunder cloud.

This phenomena is also involved, together with induction in the collection of charge from a body by a spiked conductor:

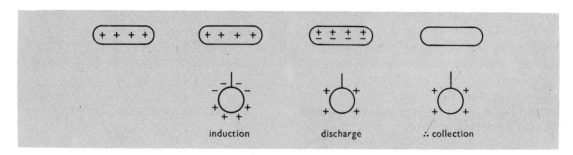

induction   discharge   ∴ collection

Fig. 203   Collection of charge by a spiked conductor

37.12   The Van de Graaff generator

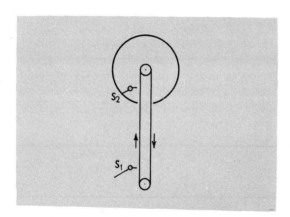

Fig. 204   The Van de Graaff generator

Charge is 'sprayed' on to a fast-moving belt at spikes S_1, collected at spikes S_2 and is stored on the smooth top sphere.

Very high voltages can be built up, depending on the size and smoothness of the upper sphere.

Energy is conserved because the motor must do work to take the charged belt up against the repulsion of the like charge on the sphere.

37.13   Lines of force

These can be used to indicate the nature of an electrostatic field, but they are not so easily plotted as magnetic lines of force. Light fibres and high voltages are needed.

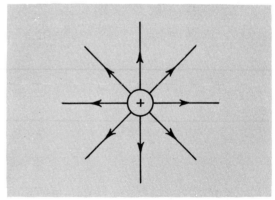

Fig. 205   Lines of force from an isolated charge

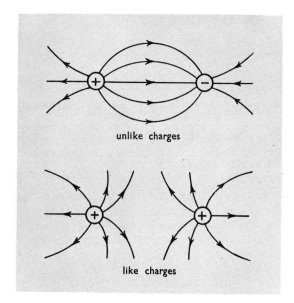

unlike charges

like charges

Fig. 206　Lines of force between two charges

38　QUANTITATIVE ELECTROSTATICS

38.1　The force between charges

This is another case of an inverse square law being applicable. The force F (newtons) between two charges Q_1, Q_2 (coulombs), distance d (metres) apart is given in vacuo by

$$F = \frac{1}{4\pi\epsilon_0} \cdot \frac{Q_1 Q_2}{d^2}$$

This can be roughly verified by using a torsion balance, but an accurate direct test is not available.

38.2　The constant ϵ_0

As the unit of charge is defined electromagnetically it can be shown experimentally that in SI units $4\pi\epsilon_0 = 1\cdot11 \times 10^{-10}$ F/m, so that $\epsilon_0 = 8\cdot85 \times 10^{-12}$ F/m. The name given to this quantity is the permittivity of free space.

38.3　Electric intensity, E

The electric intensity at a point is the force in newtons experienced by a positive charge of one coulomb placed at the point.

At a distance r (metres) from a charge $+Q$ (coulombs) this becomes

$$E = \frac{1}{4\pi\epsilon_0} \cdot \frac{Q \cdot 1}{r^2}$$

$$= \frac{Q}{4\pi\epsilon_0 r^2}$$

directed along the line from the charge $+Q$ to the test charge.

If several charges are present the resultant intensity is found by adding, vectorially, the component intensities.

38.4 Potential, V

To move a charged body against the electric intensity, will require the expenditure of work, which is stored as the potential energy of the body. At great distances from the charged objects which are causing the electrostatic field, the electric intensity will be effectively zero, and the moving charged body will have no potential energy.

The potential energy of a positive charge of one coulomb placed at any point in an electric field is called the electric potential at that point. It is defined as the work in joules needed to bring a positive charge of one coulomb from infinity to that point against the electric intensity.

The difference in potential (p.d.) between two points is therefore the work done in joules in taking a positive charge of one coulomb from one point to the other against the electric intensity.

The potential at a point is one volt if one joule of work is done on, or by, a positive charge of one coulomb in moving from infinity to that point.

38.5 The potential due to a single charge

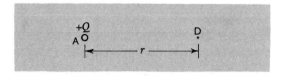

Fig. 207 Potential due to a point charge

To find the potential at D, consider a positive charge of one coulomb at some point C, distance $x \, (> r)$ from A, in vacuo.

Then the intensity at C, as proved above,

$$= E_x = \frac{Q}{4\pi\epsilon_0 x^2}$$

If the charge moves through a distance δx the work done is $E_x.\delta x$ and so the potential at D

$$= V_D = \int_r^\infty E_x \cdot \mathrm{d}x$$

$$= \int_r^\infty \frac{Q \cdot \mathrm{d}x}{4\pi\epsilon_0 x^2}$$

$$= \left[-\frac{Q}{4\pi\epsilon_0 x} \right]_r^\infty$$

$$= \frac{Q}{4\pi\epsilon_0 r}$$

Potential is a scalar, and the potential due to several charges is the algebraic sum of the component potentials.

38.6 The relation between intensity and potential

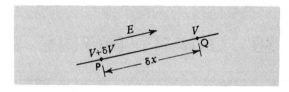

Fig. 208 Relation between potential and electric intensity

Consider two points P and Q, δx apart, with the electric intensity in the direction PQ being E. Therefore if a positive charge of one coulomb goes from P to Q the work done by the field is $E.\delta x$. This means that work has to be done to return the charge from Q to P. So P is at a higher potential than Q as shown.

Therefore in going from P to Q the change in potential is $-\delta V$

$$E\delta x = -\delta V$$

Note E and x are measured in the same sense.

So in the limit

$$E = -\frac{dV}{dx}$$

i.e. the field equals the negative gradient of potential and E is expressed in volts/metre.

38.7   The field and potential inside a conductor

In electrostatics all points in a conductor must be at the same potential or there would be a movement of charge. So V is constant, and therefore

$$E = -\frac{dV}{dx} = 0$$

Thus there is no field intensity inside a charged conductor

38.8   A particle between charged plates

Consider a particle carrying a charge of e (coulombs) between two plates, one of which is at a potential V(volts,) while the other is earthed.

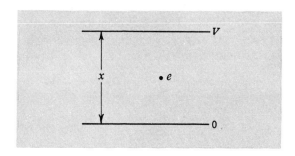

Fig. 209   Particle between charged plates

If the field intensity $= E$
∴   the force on the charge $= eE$
If the field is uniform

$$E = \frac{V}{x}\left(\text{cf. } \frac{dV}{dx} = -E\right)$$

so

$$\text{force} = \frac{eV}{x}$$

and

$$\text{the acceleration} = \frac{eV}{mx}$$

where $m =$ the mass of the particle in kilograms

38.9   Capacitance

An isolated charged object must have a potential due to its own charge, but for a given charge the potential depends on the size and shape of the object.

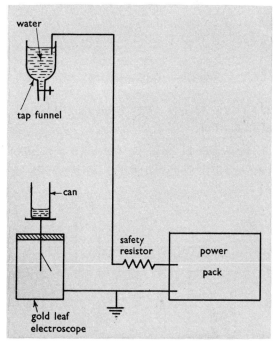

Fig. 210   The concept of capacitance

Each equal drop carries an equal charge. The bigger the can the larger the number of drops needed for the same deflection, that is the same potential. Thus the larger can has the larger capacitance for charge.

It is found that charge is proportional to potential and so

$$Q \propto V$$

and

$$Q = CV$$

where $C =$ the capacitance of the object concerned. It is defined as the charge necessary to raise the potential by unity.

The potential of a charged conductor can be lowered by the proximity of another conductor. Such a system of conductors is called a capacitor, the capacitance of which equals the charge on either conductor divided by the potential difference between them.

The unit of capacitance is the farad, which is the capacitance of a conductor the potential of which rises by one volt when it receives a charge of one coulomb. The farad is a very large unit so the following are in common use

$$1 \text{ microfarad} = 1 \ \mu\text{F} = 10^{-6} \text{ F}$$

$$1 \text{ picofarad} = 1 \text{ pF} = 10^{-12} \text{ F}$$

38.10   The factors which determine the value of a capacitor

These can be investigated with a parallel plate capacitor and a gold leaf electroscope.

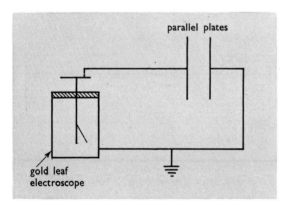

Fig. 211   The parallel plate capacitor

The capacitor is charged and its potential is indicated by the electroscope.

(*i*)   If the plates are moved apart the leaf moves further out. This indicates an increase in potential at constant charge, so the capacitance has decreased.

(*ii*)   A sheet of dielectric is inserted between the plates, the leaf diverges less, so there is a higher capacitance.

(*iii*)   The area of the plates is increased,

the leaf diverges less, so there is a higher capacitance.

Actually $$C = \frac{\epsilon_0 A}{d}$$

in vacuo, ignoring edge effects.

38.11   Capacitors in series and parallel

(*a*)   *Parallel*

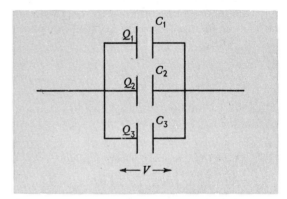

Fig. 212   Capacitors in parallel

The potential across the capacitors is the same for all, V. The equivalent capacitor C, will store charge Q ($= Q_1 + Q_2 + Q_3$) for the same potential V.

$$\therefore \quad Q_1 = C_1 V \quad Q_2 = C_2 V \quad Q_3 = C_3 V$$

and $$Q = CV$$

but $$Q = Q_1 + Q_2 + Q_3$$

$$\therefore \quad CV = C_1 V + C_2 V + C_3 V$$

$$C = C_1 + C_2 + C_3.$$

(*b*)   *Series*

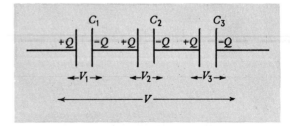

Fig. 213   Capacitors in series

The charge on each capacitor is the same. The equivalent capacitor will store the same charge under a potential $V\,(=V_1+V_2+V_3)$

$$\therefore\quad V_1=\frac{Q}{C_1}\quad V_2=\frac{Q}{C_2}\quad V_3=\frac{Q}{C_3}$$

and

$$V=\frac{Q}{C}$$

but

$$V=V_1+V_2+V_3$$

$$\therefore\quad \frac{Q}{C}=\frac{Q}{C_1}+\frac{Q}{C_2}+\frac{Q}{C_3}$$

$$\therefore\quad \frac{1}{C}=\frac{1}{C_1}+\frac{1}{C_2}+\frac{1}{C_3}$$

38.12   The energy of a charged capacitor

Suppose that a capacitor of capacitance C is charged to a final potential V, storing charge Q.

During the charging, suppose that at some instant the charge and potential are q and v respectively. Now let a further charge δq be brought up from infinity,

work done $\delta W=v.\delta q$

by definition of potential

but $v=q/C$

$$\therefore\quad \delta W=\frac{q.\delta q}{C}$$

and in the limit $dW=\frac{q.dq}{C}$

and integrating from the uncharged to the fully charged condition, the work done in charging, W

$$=\int_0^Q \frac{q\,dq}{C}=\frac{Q^2}{2C}=\tfrac{1}{2}CV^2=\tfrac{1}{2}QV.$$

and this is the energy stored in the capacitor. This energy can be released as a spark, or can drive an electric motor for a short time, etc.

If C is in farads, Q in coulombs, V in volts, then W will be in joules.

38.13   The comparison of capacitances

(a)   Calibrated electroscope

Known voltages are applied to calibrate the scale of the electroscope.

The electroscope (capacitance C) is charged to a known voltage V_1, a known capacitance C_1 connected in parallel with it, and the new voltage V_2 is found.

Assuming no loss of charge,

$$\text{Charge}=CV_1=(C+C_1)\,V_2$$

Therefore C can be found and the process extended to other capacitances.

(b)   By switching

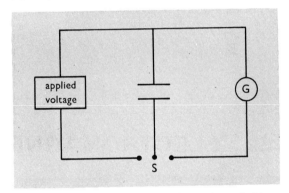

Fig. 214   Measurement of capacitance by switching

The switch S, worked from a known frequency (e.g. the a.c. mains) allows the capacitor to charge and discharge n times a second.

$$\text{Charge}=CV$$

where $C=$ capacitance
          $V=$ applied voltage

$\therefore$ Charge/sec $=CVn=$ current measured.

If the capacitor is made from two sheets of glass of known area with their inner surfaces coated with 'Aquadag', and separated at the corners and centre by small perspex spacers of known thickness, then the measured capacitance may be compared to the formula

$$C = \frac{\epsilon_0 A}{d}$$

and an experimental value of ϵ_0 determined.

(c) *By ballistic galvanometer* See §39.6.

38.14 The charging of capacitors

A capacitor takes a finite time to charge or discharge depending on the capacitance and resistance in the circuit.

The time constant of the circuit ($C \times R$) is a measure of the charging (and discharging) time. Actually it is the time required to reach a definite fraction (63·2%) of the final voltage or charge. A capacitor connected to a battery with ordinary connecting wire has an almost zero time constant.

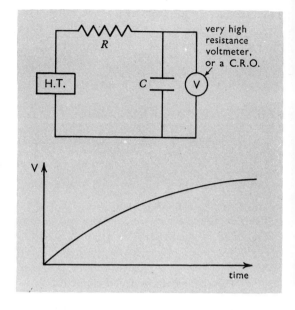

Fig. 215   The charging of a capacitor

39   ELECTROMAGNETISM

39.1 The basic formulae

The magnetic field created by a current will exert a force on a magnetic pole near to it. Likewise the magnet will exert a force on the current, as will any magnetic field.

The existence of this force can be shown by running a wire between two strong bar magnets. When a current is passed the wire will move up or down depending on the relative directions of the field and the current.

The direction can be worked out by Fleming's Left hand rule:

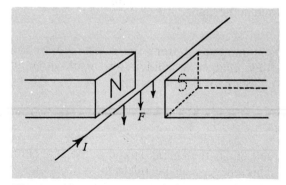

Fig. 216   Force on a current carrying conductor in a magnetic field

Hold the left hand with thumb, first and second fingers mutually perpendicular. If the first finger indicates the field and the second finger the current, then the thumb points in the direction of the force.

Various 'toys' like Barlow's wheel, and Faraday's rotating wire will show the same effects.

The magnitude of the force defines the strength of the magnetic field in terms of its magnetic flux density B.

$$\text{Force} = BIl$$

where the force is measured in newtons,

where I = current in amp,

l = length of the conductor at right angles to the field, in metres,

B in tesla

$(1 \text{ tesla} = 1 \text{ Wb m}^{-2})$

If the current is considered as a series of moving charged particles, each having charge e (coulombs), and velocity v, the particles moving along a wire of length l which contains N charges,

It takes time l/v for a charged particle to move along the conductor,

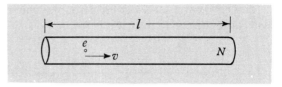

Fig. 217　Force on a moving charge in a magnetic field

∴ in this time charge Ne leaves the end of the conductor,

∴ rate of flow of charge

$$= \frac{Ne}{l/v} = \frac{Nev}{l} = \text{current}$$

∴ force $= \dfrac{B.Nev.l}{l} = BNev$

which is true whether the conductor is there or not, so the force on one charged particle

$$= Bev$$

39.2　A coil in a magnetic field

Consider a coil, pivoted as shown, at an angle θ to a uniform magnetic field.

Then the forces are as shown. The vertical forces are resisted by the rigidity of the coil and have no effect.

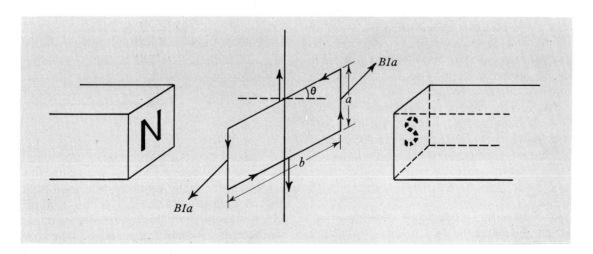

Fig. 218　Coil in a magnetic field

The horizontal forces exert a couple which will cause the coil to turn, if it is free to do so.

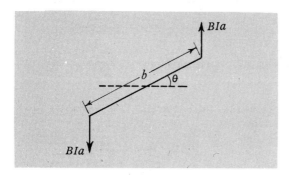

Fig. 219   Torque on coil in a magnetic field

The torque of the couple

$$= BIa.b \cos \theta \quad \text{(for one turn)}$$

and if A is the area of the coil and if there are N turns on the coil

$$\text{torque} = BIAN \cos \theta$$

so the coil turns until $\theta = 90°$, i.e. perpendicular to the field.

39.3   The moving coil galvanometer

If the movement of the coil is opposed by a couple which depends on the angle through which the coil has turned, then the position of the coil depends on the strength of the current. The instrument so formed is capable of measuring current. This is basically the moving coil galvanometer.

In order to have a uniform scale the field is made radial and not uniform, by using concave pole pieces and a centre cylindrical core. The coil is always parallel to the field so

$$\text{torque} = BIAN$$

and if restoring torque $= \tau\theta$ where $\tau =$ torsional constant of the suspension,

$$\tau\theta = BIAN \text{ and } \theta \propto I$$

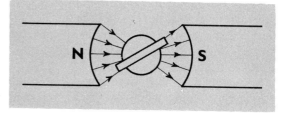

Fig. 220   Radial field for a moving coil galvanometer

The design of a moving coil galvanometer depends on the sensitivity required.

$$\text{Sensitivity} = \frac{\theta}{I} = \frac{BAN}{\tau}$$

The most sensitive have a large suspended coil so that A is large and τ is small.

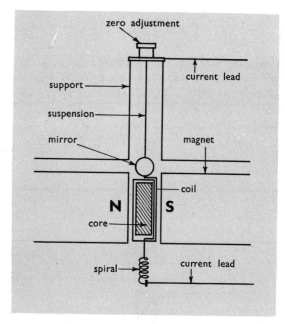

Fig. 221   A suspended coil galvanometer

The coil of many turns is suspended by a phosphor-bronze strip from a top support which usually contains a zero adjustment mechanism. The coil hangs between the concave poles of a magnet and round a cylindrical core. The phosphor bronze serves to lead the current to the coil and it is led out through a spiral fixed to the case. The movement of the coil is usually followed by shining a beam of light on to a concave mirror, the light being reflected and forming a spot on a translucent scale.

In older versions the meter and the lamp and scale were quite separate, but now they can be built into the same case.

A clamp is normally provided so that the suspension will not be damaged when the instrument is moved.

These instruments are too delicate for many purposes, therefore pivoted coil instruments, which are more robust, but less sensitive, are used. The principle is the same, but the coil is pivoted, in jewelled bearings, and controlled by two hair springs. A pointer is used, moving over a calibrated scale.

Modern developments have led to putting a cylindrical magnet inside a mild steel cylinder and using pole pieces. This creates a smaller instrument in which the magnetic field is well screened and therefore two close instruments do not interfere.

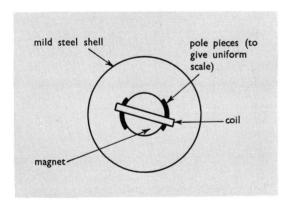

Fig. 222   Moving coil system with central magnet

The latest advance is to combine the two ideas and produce a pointer instrument where the coil is held by two short lengths of beryllium–copper ribbon under tension. The ribbon performs the functions of the pivots and of the hair springs, yet it is much more robust than the long fine suspension of the older instruments.

39.4   Moving coil ammeters

Instruments designed as above are usually too sensitive for general use and have to be modified. In order to turn a galvanometer into an ammeter a low resistance shunt is placed across the terminals.

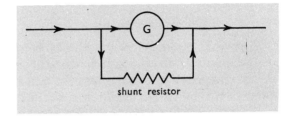

Fig. 223   Conversion of a galvanometer into an ammeter

The value of the shunt must be chosen so that most of the current flows through it. Just enough current to work the meter must take the alternative path through the meter.

An ammeter has to be put in series with the circuit in which the current has to be measured. It has, therefore, to be of low resistance, (often due mainly to the shunt) in order to affect the circuit current as little as possible.

39.5   Moving coil voltmeters

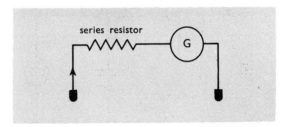

Fig. 224   Conversion of a galvanometer into a voltmeter

In this case a high resistance is placed in series with the galvanometer. When the maximum potential difference to be measured is applied, then the maximum current the meter is designed to handle must flow.

A voltmeter is always placed across the component (etc.) under test and the higher its resistance the less it affects the circuit under test. Information on the resistance of voltmeters is usually given as so many ohms per volt, 1000 ohms per volt being a moderate figure, and higher values being obtained with more sensitive instruments and then with electronic circuits.

39.6   The ballistic galvanometer

This is a sensitive type of galvanometer designed so that there is as little damping of the coil as possible. In particular the coil is wound on a non-conducting former and there is no short-circuiting shunt so that electromagnetic damping is eliminated.

Under these conditions, when an electrical impulse is given to the coil (e.g. when a capacitor is discharged through it) the coil has hardly begun to move before the impulse has passed. The electrical energy is turned into kinetic energy of the coil which twists against the suspension. The coil then oscillates with a slowly decaying amplitude.

The important thing is that the initial swing is proportional to the charge passed.

Various uses will be discussed later, but a simple example is in the comparison of the capacitances of two capacitors.

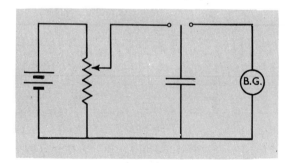

Fig. 225 A ballistic galvanometer used for the comparison of capacitors

The larger capacitor is charged to such a voltage that on discharge through the ballistic galvanometer an 'on-scale' deflection occurs. Then the capacitor is replaced by the other under test and the process repeated at the

same voltage. Lower voltages can then be used for further readings.

First capacitor $Q_1 = C_1 V = k\theta_1$

Second capacitor $Q_2 = C_2 V = k\theta_2$

$$\therefore \frac{C_1}{C_2} = \frac{\theta_1}{\theta_2}$$

where Q = charge
C = capacitance
V = p.d. applied
θ = deflection
k = galvanometer constant

39.7   The d.c. motor

Instead of opposing the turning of the coil it is very often desired to maintain a continuous rotation. This can be done if the current through the coil is reversed every half turn of the coil, by using a split-ring commutator.

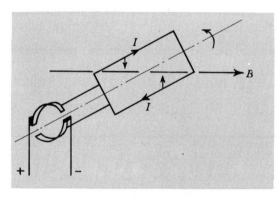

Fig. 226   The d.c. motor.

39.8   A magnet in a magnetic field

It has been shown that a coil experiences a couple in a magnetic field. A permanent magnet will also experience a couple which is found by experiment to be given by

$$\text{torque} = MB \sin \theta$$

where B = magnetic flux density

M is a constant for the magnet, called its magnetic moment and defined as the couple required to hold the magnet at right angles to a field of 1 tesla.

If à magnet is allowed to come to equilibrium in two mutually perpendicular fields then the fields are related by

$$B_2 = B_1 \tan \theta$$

where B_1, B_2 = magnetic flux density of the fields

θ = angle between B_1 and the magnet

In particular, in deflection magnetometers B_1 is provided by the earth's magnetic field so

$$B = B_{\text{earth}} \tan \theta$$

39.9   The Ampere/Laplace formula

There is always a magnetic field associated with the flow of current along a conductor, and the field depends upon the shape and position of the conductor as well as the strength of the current.

The formulae for the various cases can be worked out from the Ampere/Laplace theory which refers to an infinitesimal current element.

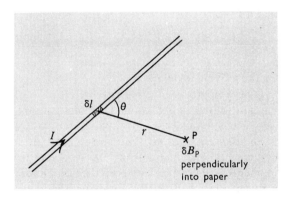

Fig. 227   The Ampere/Laplace equation

The field at P (δB_P) due to the current I flowing through the element δl is given by

$$\delta B_P = \frac{\mu_0}{4\pi} \cdot \frac{I.\delta l.\sin \theta}{r^2}$$

This equation and the ones derived from it are true only in the absence of magnetic materials. $\mu_0 = 4\pi \times 10^{-7}$, and has this value to avoid changing the value of the ampere.

39.10   Important cases

(*a*)   *The field at the centre of a circular coil*

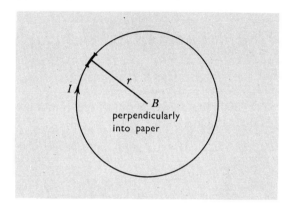

Fig. 228   Field at the centre of a circular coil

$$B = \frac{\mu_0}{4\pi} \cdot \frac{2\pi NI}{r} = \frac{\mu_0}{2} \cdot \frac{NI}{r}$$

where N = no. of turns

I = current

r = radius

This can be verified with a varicoil magnetometer. This consists of a deflection magnetometer mounted at the centre of a coil system so arranged that N, r and I can all be varied one by one.

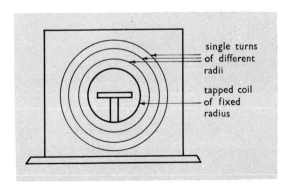

Fig. 229   The varicoil magnetometer

With a fixed current and one turn it can be shown that

$$B = B_{earth} \tan \theta \propto \frac{1}{r}$$

With a fixed radius and current it can be shown that

$$B \propto N$$

With a fixed radius and number of turns it can be shown that

$$B \propto I$$

Note that the magnetometer coil must be placed in the meridian.

(b)   *The field due to a long straight wire*

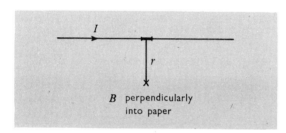

Fig. 230   Field due to a long straight wire

$$B = \frac{\mu_0}{4\pi} \cdot \frac{2I}{r}$$

This can be verified by supporting a long straight wire in the magnetic meridian above a deflection magnetometer. With a fixed current the distance from the wire to the magnetometer is varied, and with a fixed distance the current is varied.

(c)   *The field inside a long solenoid*

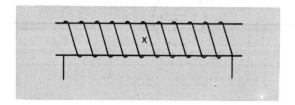

Fig. 231   Field at the centre of a solenoid

$$B = \frac{\mu_0}{4\pi} \cdot 4\pi nI = \mu_0 \cdot nI$$

where n = turns per metre

This is strictly true only for an infinitely long solenoid but providing the length is about ten times the diameter the field in the centre is uniform and given by the above equation. The field remains constant over the central cross-section.

This can be verified by using extensible solenoids. These are solenoids wound without a former so that they can be extended. A small magnetometer is mounted at the centre so that the usual measurements can be made.

39.11   The definition of the ampere

For an infinitely long wire, in vacuo, carrying a current I the field, at a distance r from it, is

$$\frac{\mu_0}{4\pi} \cdot \frac{2I}{r}$$

so for a parallel wire at this distance carrying a current I', the force per unit length on it is

$$\frac{F}{l} = \frac{\mu_0}{4\pi} \cdot \frac{2I}{r} \cdot I'$$

and if $I = I'$

$$\frac{F}{l} = \frac{\mu_0}{2\pi} \cdot \frac{I^2}{r}$$

So the ampere is defined:

'The ampere is the intensity of a constant current which, if maintained in two parallel, rectilinear conductors of infinite length, of negligible circular section and placed at a distance of one metre from one to the other in vacuo, will produce between the conductors a force equal to 2×10^{-7} newtons per metre of length.' This is consistent with $\mu_0 = 4\pi \times 10^{-7}$.

39.12   The current balance

This definition of the ampere enables current to be measured in terms of the force between two conductors, not only when they are parallel wires, but also when other geometrical arrangements are used.

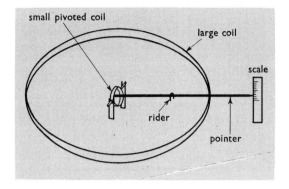

Fig. 232   A current balance

One possibility is to have a large coil (N turns, radius R) providing a field at the centre where a second small coil (n turns radius r) is pivoted as shown. The pivots are also the current leads into the small coil. When a current I flows through both coils the field at the centre of the larger coil is

$$B = \frac{\mu_0 NI}{2R}$$

and so the couple on the smaller coil is

$$B \cdot \pi r^2 nI = \frac{\mu_0 \cdot \pi r^2 n NI^2}{2R}$$

which is balanced against the extra couple due to moving the rider of weight mg(newtons) a distance d (metres) to restore the pointer to its original position.

$$mgd = \frac{\mu_0 \pi r^2 n NI^2}{2R}$$

40   CURRENT, CHARGE, POTENTIAL DIFFERENCE, POWER

40.1   Current

This has already been defined from its magnetic effect (see §**39.11**)

40.2   Charge

The unit of electric charge, called the coulomb is the quantity of electricity transported in one second by a current of one ampere.

40.3   Potential Difference   (*c.f.* §**38.4**)

One volt is the difference in potential between two points of a conducting wire carrying a constant current of one ampere, when the power dissipated between these points is equal to one watt.

40.4   Power

From the above definitions it is clear that work is needed to drive a current round a circuit and therefore this current will release energy as it goes round the circuit. In particular if a current of I (amperes) flows under a potential difference of V (volts) the work released is IV (joules s^{-1} or watts.) usually in the form of heat

40.5   The rating of electrical devices and the cost of using them

Most household devices are rated in watts (or kilowatts, 1 kilowatt = 1000 watts). When they are used for several hours the energy used = kilowatts × hours (in kilowatt-hours).

1 kilowatt-hour (=3,600,000 joules) is called the (Board of Trade) Unit and commercial electricity is charged for at so much per unit.

40.6   The interconversion of electrical energy into other forms

Note that electrical energy can also be converted into:

light	(bulb)
sound	(loudspeaker)
chemical	(charging accumulators)
mechanical	(motor)

but these are not easily treated in a quantitative way, especially as heat is usually produced as well.

40.7   The variation of current with p.d. in different devices

If the current flowing through a given device is measured for different applied potential differences the resulting graphs can vary widely. Most cases will be dealt with in context but a convenient list for cross reference is given here.

(*i*)   Discharge tubes (§**47.1**).

(*ii*)   Valves (§**48.2, 48.5**).

(*iii*)   Liquids undergoing electrolysis (§**43.4**).

(*iv*)   Non-linear resistors (Fig. 233).

(*v*)   Metallic conductors (§**40.8**)

40.8   Metallic conductors

These are a very important group where, as was first found by Ohm:

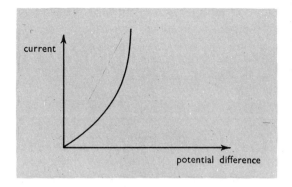

Fig. 233   Current variation with p.d. for a non-linear resistor

'The current flowing through a metallic conductor maintained under constant physical conditions is directly proportional to the potential difference between its ends' (Ohm's law).

Symbolically

$$V \propto I$$

$$\therefore \quad V = \text{constant} \times I$$

and this constant is called the resistance of the metallic conductor, symbol R.

Thus

$$V = IR$$

One ohm is the resistance between two points of a conductor when a constant difference of potential of one volt applied between these two points, produces in this conductor a current of one ampere, this conductor not being the source of any electromotive force.

40.9   The verification of Ohm's law

This is not easy because most electrical instruments are calibrated by assuming Ohm's law. Absolute instruments such as a current balance and an electrostatic voltmeter must be used, where measurements may be derived from first principles.

40.10   Some applications of Ohm's law

(a)   Resistors in series

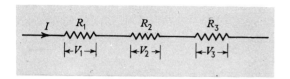

Fig. 234   Resistors in series

If a current I flows through three resistors as shown, there will be potential difference across each resistor. The equivalent resistor R must be such that when the same current I flows the potential difference across the resistor is $V = V_1 + V_2 + V_3$.

From Ohm's law

$$V = IR \quad V_1 = IR_1 \quad V_2 = IR_2 \quad V_3 = IR_3$$

$$\therefore \quad IR = IR_1 + IR_2 + IR_3$$

$$\therefore \quad R = R_1 + R_2 + R_3$$

(b)   Resistors in parallel

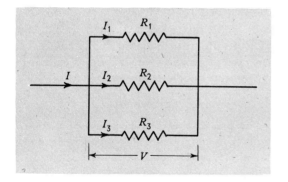

Fig. 235   Resistors in parallel

In this case the potential difference across each resistor is the same, and the equivalent resistor R must pass the same total current under this potential difference.

Note that $I = I_1 + I_2 + I_3$ as current is conserved at any junction in a circuit.

$$\therefore \quad I = \frac{V}{R} \quad I_1 = \frac{V}{R_1} \quad I_2 = \frac{V}{R_2} \quad I_3 = \frac{V}{R_3}$$

$$\frac{V}{R} = \frac{V}{R_1} + \frac{V}{R_2} + \frac{V}{R_3}$$

$$\therefore \quad \frac{1}{R} = \frac{1}{R_1} + \frac{1}{R_2} + \frac{1}{R_3}$$

(c)   Resistance boxes

Combinations of resistors which can be altered by switching or plugging are very useful in many experiments. Care should be taken not to overload resistance boxes, which, unlike rheostats, can only carry small currents.

(d)   The complete circuit

Consider a circuit consisting of a cell and resistor.

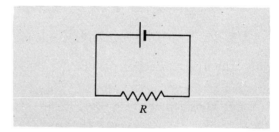

Fig. 236   The characteristics of a cell

A voltmeter which has no effect on the circuit is connected across the cell. As the resistance increases so does the reading on this voltmeter. When the resistance is infinity the reading is called the electromotive force (e.m.f.) of the cell (its maximum potential difference) otherwise it is simply the (available) potential difference.

An ammeter connected in the circuit would show that as the resistance increased the current gets less, and is zero when the resistance is infinity.

Thus the e.m.f. of a cell is the potential difference between its terminals when no current flows.

These effects arise because the cell itself has resistance, called its internal resistance. If a current flows, some of the cell's e.m.f. must be used to drive the current through the cell itself, the remainder being 'available' for the rest of the circuit.

$$\text{e.m.f.} = \text{Current} \times \text{Total resistance}$$

$$\text{p.d.} = \text{Current} \times \text{External resistance.}$$

If e.m.f. $= E$, p.d. $= V$, current $= I$. External resistance $= R$, Internal resistance $= r$

$$E = I(R+r) \quad V = IR$$
$$\therefore \quad E - V = Ir$$

N.B.

(*i*)   A cell can be thought of as:

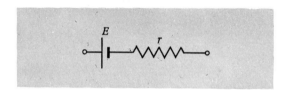

Fig. 237   The internal resistance of a cell

with no internal connection possible.

(*ii*)   For a Leclanché cell $r \simeq 5$ ohms so V varies considerably with current, but the short circuit current is not too high.

(*iii*)   For an accumulator $r \simeq 0.01$ ohms so V varies little with current, but the short circuit current can be very high.

(*e*)   *Power in a pure resistance*

As shown above

$$P = IV$$

$$\therefore \quad P = I^2R = \frac{V^2}{R}$$

where $P =$ power (watt)
$I =$ current (amp)
$V =$ p.d. (volt)
$R =$ resistance (ohm)

which are often useful forms of the equation. N.B. If a cell (etc.) of internal resistance r is feeding current into an external resistance R, the maximum power is dissipated when $R = r$.

(*f*)   *Resistivity*

Many resistors are made from lengths of wire. It can be shown that the resistance of a wire is proportional to its length and inversely proportional to its cross-sectional area.

$$\therefore \quad R = \frac{\rho l}{A}$$

where $R =$ resistance of wire
$l =$ length of wire
$A =$ cross-sectional area of wire

and ρ is a constant, the resistivity or specific resistance depending on the material of the wire, defined as the resistance of a specimen of a material 1 m in length and 1 m² in cross-sectional area (units: ohm m).

(*g*)   *Temperature coefficient of resistance*

The resistance of most metals varies with temperature. Over the temperatures usually met with in the laboratory the variation is fairly linear. Results can usually be expressed in the form:

$$R_t = R_0(1 + \alpha t)$$

where R_0 is the resistance of the specimen at 0°C, R_t the resistance at t°C and α is a constant depending on material called its temperature coefficient of resistance.

(*h*)   *Resistance thermometers*

The property of change of resistance with temperature can be used to measure temperature. The results can be given either on a platinum resistance scale (if the metal used is platinum, and the variation is assumed to be linear).

$$t = \frac{R_t - R_0}{R_{100} - R_0} \times 100$$

or on the gas scale

$$R = R_0(1 + \alpha\theta + \beta\theta^2),$$

where α and β are constants found by calibration at 0°C, 100°C and 444·6°C (b.p. of sulphur).

As discussed under thermometry t and θ may be different numbers describing the same temperature.

41 THERMOELECTRIC THERMOMETRY

41.1   The Seebeck effect

If a wire of one material is connected by wires of another material to a galvanometer and the two junctions kept at different temperatures a p.d. is set up and a current flows. The magnitudes of p.d. and current depend on the temperature difference and the materials used.

This device is called a thermocouple and the phenomenon the thermo-electric or Seebeck effect.

If the temperature is varied widely the e.m.f. varies as shown in the graph, where one junction is being kept at 0°C.

Temperature A is called the neutral temperature ($\approx$285°C for a copper–iron thermocouple) and temperature B is called the inversion temperature ($\approx$570°C for a copper–iron thermocouple).

Over a wide range of temperature, the value of the e.m.f. E_θ with the hot junction at θ°C (gas scale) and the cold junction at 0°C is given by:

$$E_\theta = \alpha\theta + \beta\theta^2$$

where α and β are constants.

The most important use of thermocouples is for temperature measurement (low temperatures, lead–gold; medium temperatures copper–iron, etc., high temperatures, platinum–platinum and rhodium alloy or platinum and iridium alloy).

The direction of the current flow at the cold junction is from the first to the second in this list: Antimony, iron, gold, (zinc, copper), lead, platinum, constantan, bismuth.

The Seebeck effect is not the only thermoelectric effect; others called the Peltier and Thomson effects are also known to exist.

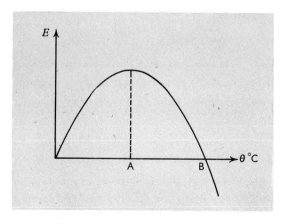

Fig. 238   E.m.f. of a thermocouple

42 THE POTENTIOMETER AND THE WHEATSTONE NETWORK

42.1 The theory of the potentiometer

This instrument is designed to measure potential differences. In its simplest form it consists of a long uniform resistance wire, connected to a 'driver' cell, i.e. one capable of maintaining a steady e.m.f. over a long period (usually a lead accumulator).

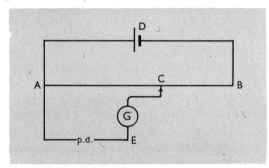

Fig. 239 The basic potentiometer circuit

AB is the resistance wire, D the driver cell. A movable contact C is slid along AB until there is no deflection of the galvanometer, when the p.d. to be measured is connected as shown.

Then p.d. between E and C is 0 at balance

$\therefore$ p.d. between A and E
$$= \text{p.d. between A and C.}$$

Now if the wire has resistance/m, ρ, and a current I flows, then, as no current flows through the galvanometer

p.d. across $AB = \rho I \, AB$

p.d. across $AC = \rho I \, AC$

$\therefore \quad \dfrac{\text{p.d. across AC}}{\text{p.d. across AB}} = \dfrac{\rho I \, AC}{\rho I \, AB} = \dfrac{AC}{AB} = \dfrac{\text{p.d. across AE}}{\text{p.d. across AB}}$

or
$$\text{p.d. across AE} = AC \cdot \dfrac{\text{p.d. across AB}}{AB}$$

In practice the p.d. to be measured can take various forms, but the same basic ideas apply throughout.

42.2 The comparison of the e.m.f.'s of two cells

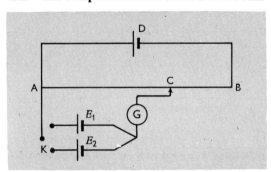

Fig. 240 Comparison of e.m.f.s

By using a two-way switch, K, either cell can be connected in circuit so that the balance position can be found at C_1 for E_1 and C_2 for E_2. Since cells E_1 and E_2 are supplying no current, the p.d. across their terminals equals their e.m.f.

Then
$$\frac{E_1}{E_2} = \frac{AC_1}{AC_2}$$

42.3 Practical points for this and other uses

(*i*) A switch in ADB can be used to reduce current drain on D when it is not in use.

(*ii*) A safety resistance should be used to protect the galvanometer.

(*iii*) If a variable resistor is placed in ADB a set of readings can be obtained. A balance reading must be obtained for each cell before the rheostat is adjusted.

(*iv*)   If C is brought near to A and then B, deflections should be in opposite directions, if not suspect that

    (*i*)   *E* the wrong way round, or
    (*ii*)   *E* is greater than the e.m.f. of D.
    (*iii*)   There is a broken connection.
    (*iv*)   The driver cell is run down.
    (*v*)   A broken galvanometer.

42.4   The measurement of internal resistance

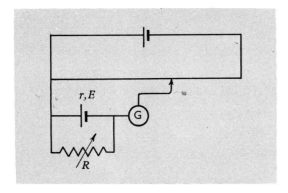

Fig. 241   Measurement of internal resistance

The balance point is found for various known values of *R*. As little current as possible is taken from the cell to minimize polarization.

Let *l* be the balance length when $R = \infty$
Let *l'* be the balance length at *R*
Then if the e.m.f. of the cell is *E* and the p.d., when connected to *R*, is *V*
   Then

$$\frac{V}{E} = \frac{l'}{l}$$

but $\qquad\qquad E - V = Ir$

where *I* is current flowing through *R* (see §**40.10d**).

and $\qquad\qquad I = \dfrac{V}{R}$

$$\therefore \quad E - V = \frac{Vr}{R}$$

$$V = \frac{E}{[1 + (r/R)]}$$

$$\therefore \quad \frac{E}{V} = 1 + \frac{r}{R} = \frac{l}{l'}$$

$$\therefore \quad r = R\left(\frac{l}{l'} - 1\right) \quad \text{or} \quad \frac{1}{l'} = \frac{1}{R} \cdot \frac{r}{l} + \frac{1}{l}$$

so calculate *r* or plot $1/l'$ against $1/R$ when the gradient is r/l

42.5   The measurement of a very small e.m.f. (e.g. *from a thermocouple*)

The potentiometer has to be calibrated, using a standard cell (i.e. one whose e.m.f. is very accurately known), and a sensitive galvanometer is needed.

   (*i*)

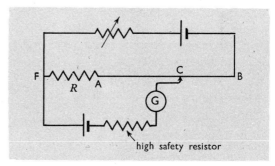

Fig. 242   Calibration of potentiometer for use with a thermocouple

A balance point is found with $AC = l_1$. Then if current flowing through FB is *I*, and the resistance per m of AB is ρ then $E_s = I(R + l_1\rho)$ where E_s = e.m.f. of standard cell.

   (*ii*)

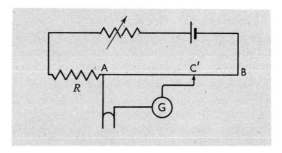

Fig. 243   Measurement of e.m.f. of a thermocouple

With the driver part of circuit untouched, let the balance be at C',

where $\qquad\qquad AC' = l_2$.

$$E_T = Il_2\rho$$

where E_T = e.m.f. of thermocouple

$$\therefore \quad \frac{E_T}{E_s} = \frac{l_2\rho}{R + l_1\rho}$$

so if ρ is found E_T can be measured.

42.6 The measurement of large currents

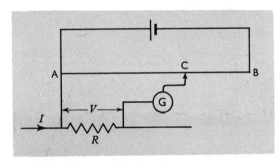

Fig. 244   Measurement of a large current

As $V = IR$, V, the potential difference, developed is compared with that of a standard cell. R is known, so I can be found. By altering the current in the driver circuit with a rheostat, the wire can be calibrated with the standard cell, so that the instrument becomes direct reading (e.g. 50 cm is equivalent to 1 volt.)

42.7 The comparison of resistances

The two resistances are connected in series with a variable resistance and an accumulator, so that a steady current (I) is sent through them. Potential differences exist across the resistances of V_1 ($= IR_1$) and V_2 ($= IR_2$). These are compared as indicated, and if $AC = l_1$, $AC' = l_2$.

$$\frac{V_1}{V_2} = \frac{R_1}{R_2} = \frac{l_1}{l_2}$$

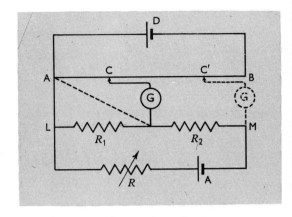

Fig. 245   Comparison of resistances

42.8 The end corrections of a potentiometer

These are due to

(*i*)   The terminals not being at the 0 and 100 (etc.) marks.

(*ii*)   Contact resistance.

(*iii*)   The resistance of the wire being changed due to a different alloy being formed on soldering or the cross-sectional area being changed.

If the errors are measured in terms of centimetres of bridge wire they may be represented as α and β.

So that the total length becomes $100 + \alpha + \beta$

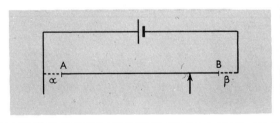

Fig. 246   The end corrections of a potentiometer

To determine α (or β)

Repeat **42.2** with E_1 (balance length l_1) with E_2 (balance length l_2), and with E_1 and E_2 in series (balance length l_3).

Then

$$E_1 = k(l_1 + \alpha)$$

$$E_2 = k(l_2 + \alpha)$$

$$E_1 + E_2 = k(l_3 + \alpha) \quad (k = \text{constant})$$

but $\quad (E_1 + E_2) = E_1 + E_2$

$$\therefore \quad k(l_3 + \alpha) = k(l_1 + \alpha) + k(l_2 + \alpha)$$

$$l_3 + \alpha = l_1 + \alpha + l_2 + \alpha$$

$$\alpha = l_3 - l_1 - l_2$$

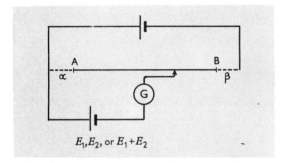

Fig. 247 Measurement of the end corrections

Repeat from the other end for β

42.9   The Wheatstone network

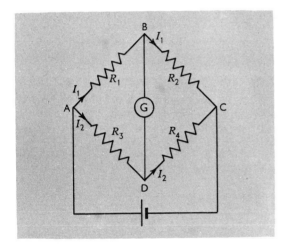

Fig. 248   The Wheatstone network

Four resistances are connected with a battery and a galvanometer as shown, and

adjusted so that there is no deflection of the galvanometer.

Then a current I_1 flows through R_1 and R_2, and a current I_2 flows through R_3 and R_4.

If no current flows through G, the potential at B and D is the same, and the p.d. between A and B equals that between A and D.

$$\therefore \ I_1 R_1 = I_2 R_3$$

Similarly $\qquad I_1 R_2 = I_2 R_4$

$$\therefore \ \frac{R_1}{R_2} = \frac{R_3}{R_4}$$

Thus if one resistance is known, and the ratio of two more adjacent resistances, the fourth can be found.

N.B. The value of the e.m.f. does not matter.

42.10   The metre bridge

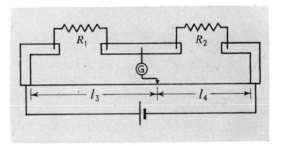

Fig. 249   The metre bridge

A metre of resistance wire is stretched between two thick copper strips, separated from a third by two gaps. Resistances R_1 and R_2 can be inserted in these gaps. A battery is connected across the wire, and a galvanometer joined between R_1 and R_2, with its other connection somewhere on the wire. When there is no deflection of the galvanometer

Then

$$\frac{R_1}{R_2} = \frac{\rho l_3}{\rho l_4} = \frac{l_3}{l_4}$$

where ρ is resistance/m of the wire

R_1 and R_2 should be reversed and the average value for R_2 (assuming R_1 is known) found.

K

Practical points

(*i*)   A Leclanché cell is suitable.

(*ii*)   The balance point should be in the centre part of the wire.

(*iii*)   The resistance of the connecting wire is important between R_1, R_2 and the bridge, but not elsewhere.

(*iv*)   Another form of this device is the Post Office Box, where three of the four resistors are built in. One is fully variable, the other two just ratio arms.

43   ELECTROLYSIS

43.1   The basic ideas

Electrolysis occurs when the passage of an electric current, through a liquid conductor, results in chemical decomposition. The conductor is termed an electrolyte, and is usually a solution of a salt or a molten salt. The terminals by which the current enters and leaves the electrolyte are the electrodes, the (conventional) current entering at the anode and leaving at the cathode.

The nature of the product formed depends on:

(*i*)   The electrolyte $\left.\begin{matrix} \\ \\ \end{matrix}\right\{$ These govern the chemical nature; details will be

(*ii*)   The electrodes found in any chemistry text book.

(*iii*)   The current density. This governs the physical nature of the product, e.g. with copper sulphate solution and copper electrodes, a high current density gives a coarse, loose deposit.

43.2   Faraday's laws of electrolysis

(*i*)   The mass of any substance liberated during electrolysis is proportional to the product of the current and the time for which it flows; that is to the total quantity of electricity which has passed.

(*ii*)   The masses of different substances liberated by the same quantity of electricity are proportional to their chemical equivalent weights.

The laws are verified

(*i*)   By weighing the deposits in a given voltameter for various quantities of electricity.

(*ii*)   By having several different voltameters in series and finding the masses of different substances liberated.

If m is the mass liberated by a current I in time t and E is the chemical equivalent of the substance concerned

$$m \propto It \quad \text{First law}$$

$$m \propto E \quad \text{Second law}$$

$$\therefore \quad m \propto EIt$$

$$= kEIt$$

$$= eIt$$

where e is the electrochemical equivalent.

The electrochemical equivalent is, therefore, the mass of substance deposited or liberated by one coulomb of electricity.

43.3 The determination of the electrochemical equivalent

A voltameter is used with a measured current for a known time, and the mass of the product is found. The results are substituted into the above equation.

(a) *Copper*

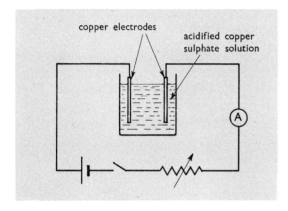

copper electrodes

acidified copper sulphate solution

Fig. 250 Measurement of the e.c.e. of copper

The circuit is set up and the current value adjusted. The cathode is cleaned mechanically and chemically, and then weighed. The circuit is reconnected, and the current allowed to flow for about half an hour. The cathode is carefully washed and dried so as not to oxidize the newly deposited copper. The cathode is then re-weighed.

(b) *Hydrogen*

A Hoffman voltameter is used. The volume is measured carefully and corrected to the value at s.t.p.

43.4 The current–voltage relationships in a voltameter

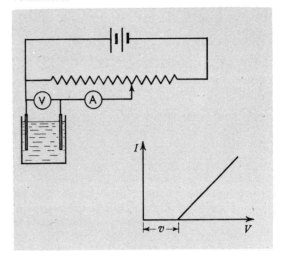

Fig. 251 Current–voltage relationship in a voltameter

With a high-resistance voltameter and a potentiometer-type arrangement the result shown can be obtained. The graph is a straight line once the voltage v, the back e.m.f. of the voltameter, is exceeded.

$$\therefore \quad I \propto (V - v)$$

v depends on the voltameter, and indicates the energy needed to cause the chemical action.

e.g. $v = 0$ volts for copper electrodes in copper sulphate solution.

$v = 1\cdot7$ volts for platinum electrodes in acidulated water.

43.5 The faraday constant

The faraday constant is the amount of electricity required to liberate one mole of a substance, and is equal to 96 487 coulombs.

44 CELLS

44.1 The simple voltaic cell

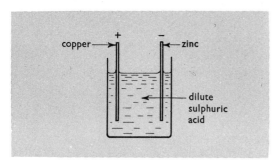

Fig. 252   The voltaic cell

The e.m.f. is approximately 1 volt, but it suffers from polarization and local action.

44.2 The Daniell cell

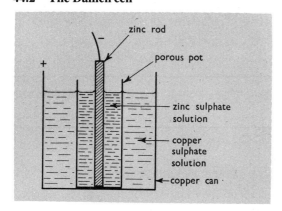

Fig. 253   The Daniell cell

The e.m.f. is 1·08 volts.

44.3 The Leclanché cell

The e.m.f. is approximately 1·5 volts.
This cell is suitable for intermittent use over long periods of time without attention.

It is also widely used in the dry cell form.

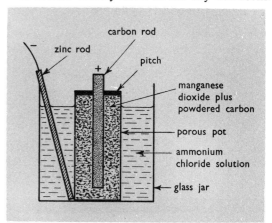

Fig. 254   The Leclanché cell

44.4 The Mallory cell

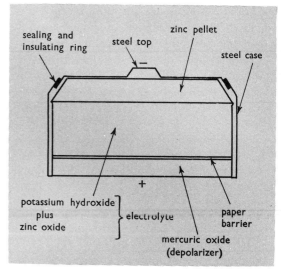

Fig. 255   The Mallory cell

The e.m.f. is 1·345 volts.

44.5 The Weston standard cell

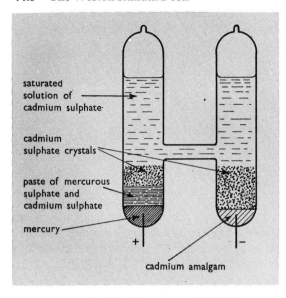

Fig. 256 The Weston standard cell

This is a standard cell in that wherever it is set up in the correct way its e.m.f. is reproduced and constant.

This e.m.f. equals 1·01859 volts at 20°C.

Only very small currents can be taken and the cell should only be used in a potentiometer circuit.

44.6 The lead accumulator

At the positive plate lead dioxide changes to lead sulphate, and at the negative plate lead changes to lead sulphate on discharge and vice versa.

The state of charge is tested by the density of the sulphuric acid.

45 ELECTROMAGNETIC INDUCTION

45.1 The basic experiments

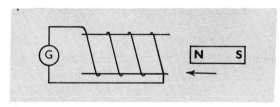

Fig. 257 Electromagnetic induction with a moving magnet

If a magnet is moved towards a coil connected to a galvanometer then there will be a deflection on the galvanometer.

An e.m.f. has been induced by the changing magnetic flux through the coil. The e.m.f. only occurs when the magnet is moving, not when it is stationary.

The size of the induced e.m.f. is found to increase if

(*i*)   the magnet is moved more quickly,

(*ii*)   a stronger magnet is used,

(*iii*)   the area of the coil is increased, or

(*iv*)   the number of turns on the coil is increased.

The direction of the induced e.m.f. in the diagram would be such as to cause a current which would create a north pole at the face of the coil nearest to the magnet as it approaches, or a south pole as it recedes.

The same effects can be demonstrated using two coils, which are electrically separate.

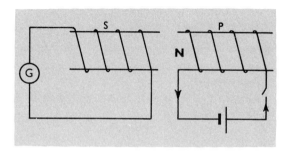

Fig. 258   Electromagnetic induction between two coils

The left-hand face of coil P becomes a N pole when current flows. As the current begins to flow, the right-hand face of coil S becomes a N pole, but as soon as the current starts to die away it will become a S pole, in both cases opposing the change that is occuring. If the current through P is constant there is no induced e.m.f. If a core of soft iron is passed through the coils the effects are greatly increased.

45.2   The laws of electromagnetic induction

The magnetic flux through an area is $\Phi = BA$ where B is the magnetic flux density (tesla) and A is the area (m²) so Φ is measured in webers. The flux linkage with a coil of N turns is $N\Phi$.

This enables the results of the observations on electromagnetic induction to be summarized in two laws:

Faraday's law

The magnitude of the induced e.m.f. is proportional to the rate of change of the total magnetic flux linked with the circuit.

Lenz's law

The direction of the induced e.m.f. is always such as to oppose the change producing it.

so
$$V = -\frac{d}{dt}(N\Phi)$$

$$= -\frac{d}{dt}(BAN)$$

where V = induced e.m.f. (volts)

The induced current = V/R but this is of greater use if V is constant over the time involved.

If a coil is connected to a ballistic galvanometer, then it is found that a definite charge circulates when the flux through the coil changes, a property which can be used to measure Φ or B.

45.3   The use of a search coil

A suitable coil (of small area and moderate number of turns) is held in the magnetic field to be measured and then moved to a region of negligible field.

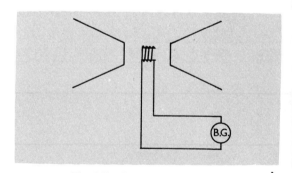

Fig. 259   Use of a search coil

now
$$V = \frac{\delta(BAN)}{\delta t}$$

so
$$I = \frac{\delta(BAN)}{R\delta t}$$

$$\therefore \quad I\delta t = \frac{\delta(BAN)}{R} = \delta Q$$

$$\therefore \quad Q = \frac{\Delta(BAN)}{R}$$

where $V =$ induced e.m.f.

　$B =$ magnetic flux density

　$A =$ area $\left.\right\}$ of search coil
　$N =$ turns

　$R =$ resistance of circuit

　$\Delta =$ total change

　$Q =$ induced charge

and if the ballistic galvanometer is calibrated Q is known and BAN can be found assuming that the final value is zero.

From this Φ or B can be deduced.

45.4　The absolute determination of resistance

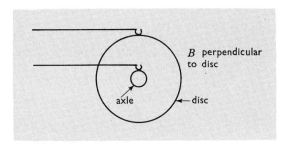

Fig. 260　Disc rotating in a magnetic field

Consider first a disc rotating perpendicularly to a uniform magnetic field with contacts at rim and axle.

Then　　　　　$V = BAf$

where　$A =$ area of disc $-$ area of axle

　　　$f =$ number of revolutions per second

Now consider this occuring as shown in a solenoid, with the solenoid current passing through a special low resistance R$'$.

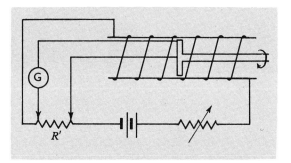

Fig. 261　Absolute determination of resistance

The circuit can be balanced so that the induced e.m.f. is equal to the potential drop across the centre part of R$'$ (of value R).

Then　　　　　　　$RI = BAf$

but　　　　　　　　$B = \mu_0 nI$

　　　　　$\therefore\ \ RI = \mu_0 nIAf$

　　　　　$\therefore\ \ \ R = \mu_0 nAf$

where $n =$ turns/metre on solenoid

Hence R may be measured without reference to other electrical quantities.

45.5　Eddy currents

Induced currents are not only set up in closed wire circuits, but occur in any conducting material, and may be troublesome if excess heating is caused. Such induced currents are referred to as eddy currents. Various illustrations are available.

(*i*)　A thick aluminium ring placed over the soft iron core of a solenoid in which alternating current is flowing will be thrown violently off, or if fixed, will become very hot.

(*ii*)　A magnetized needle is allowed to oscillate in turn over a glass sheet and then over a sheet of copper. The oscillations die away more rapidly in the second instance owing to the eddy currents induced in the metal.

45.6　The reduction of eddy currents — Waltenhofen's pendulum

If a pendulum with a 'bob' made of a metal sheet is allowed to swing between the poles of a powerful electromagnet the motion will be highly damped or even dead-beat when the magnet is switched on. However, if the 'bob' is made up of metal strips the damping is much less severe as large eddy currents are prevented. This explains the use of laminated cores in transformers.

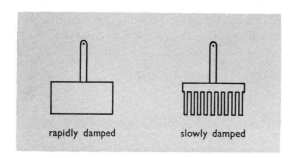

Fig. 262 Waltenhofen's pendulum

45.7 The uses of eddy currents

(*i*) Damping of sensitive balances.

(*ii*) Damping of galvanometers, the coil is wound on an aluminium former.

(*iii*) Induction furnaces.

45.8 The simple dynamo

In its simplest form the dynamo is a coil rotating in a magnetic field.

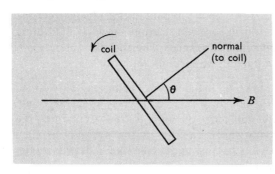

Fig. 263 The simple dynamo

Let the area of the coil $= A$, the number of turns $= N$, and the magnetic flux density $= B$. If the coil can rotate about an axis perpendicular to B then the angle of turn θ, is measured as shown.

Total flux linked with coil $= BAN \cos \theta$
Let $\theta = \omega t$
$$\therefore \quad N\Phi = BAN \cos \omega t$$
so the induced e.m.f. is

$$V = \frac{-\mathrm{d}(N\Phi)}{\mathrm{d}t} = BAN\omega \sin \omega t$$

or $\qquad V = V_0 \sin \omega t$

where $\qquad V_0 = BAN\omega$

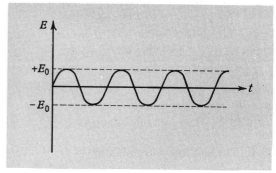

Fig. 264 E.m.f. from a simple dynamo

Note

(*i*) $V = 0$ when $\omega t = \theta = 0$, 180° etc. i.e. the coil is perpendicular to the field.

(*ii*) V is maximum when $\theta = 90°$, 270° etc. i.e. the coil is parallel to the field and there is the greatest rate of change of flux.

This e.m.f. is changing in direction and magnitude, regularly and continuously, so alternating current will flow in the circuit if it is connected by slip rings and brushes. A commutator will provide the d.c. output shown as a full line in the diagram. (Fig. 265).

In actual practice multi-coil armatures and complicated commutators are used to effectively smooth out the varying d.c.

45.9 The electric d.c. motor

If the arrangement above with a commutator is employed, but d.c. is supplied, the armature will turn. This occurs because the current-carrying conductor in a magnetic field experiences a force.

45.10 The use of a starting resistance in an electric motor

The motor is constructed similarly to the dynamo, and when it is rotating it acts as a dynamo, setting up a back e.m.f., which may

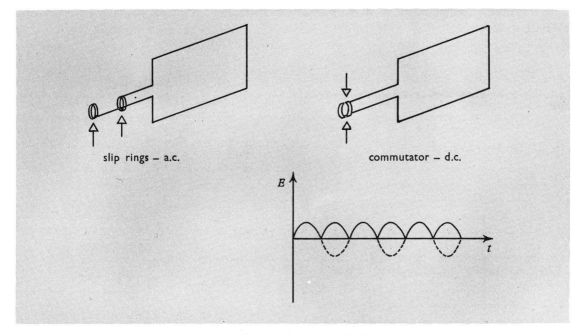

Fig. 265   Outputs from a simple dynamo

be very nearly equal to the driving e.m.f. Now if the coil has to stand the full driving e.m.f. when stationary but only a small fraction of it when running, it still has to be made to stand the full e.m.f., so adding unnecessarily to its weight.

Thus if a resistance is placed in series with the motor, and gradually decreased as the speed and therefore the back e.m.f. increases, only a thin winding is needed on the armature.

45.11   The transformer (alternating current only)

If alternating current is supplied to the primary coil, alternating current is induced in the secondary coil. Assuming no loss of magnetic flux,

$$\frac{\text{Applied voltage at primary}}{\text{Voltage at secondary}}$$

$$= \frac{\text{Number of turns on primary}}{\text{Number of turns on secondary}}$$

This is only an approximate relationship. A full treatment is very difficult. It the trans-

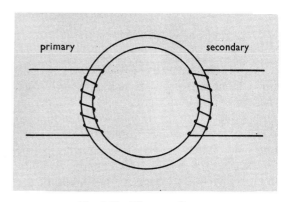

Fig. 266   The transformer

former is ideally efficient there will be no loss of energy and

$$\frac{\text{Secondary current}}{\text{Primary current}} = \frac{\text{Primary voltage}}{\text{Secondary voltage}}$$

Both these relationships are approximate. Energy losses are, however, caused by:

(*i*)   Leakage of magnetic flux.

(ii) Eddy currents—reduced by a laminated core.

(iii) Hysteresis loss—reduced by a suitable material.

(iv) Heating in the coils—according to Joule's law.

45.12 The induction coil

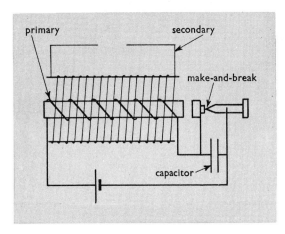

Fig. 267   The induction coil

A coil of a few turns of thick wire is wound on a laminated core. It is connected via a make-and-break to a direct current supply. A coil of many turns of thinner wire is wound on the core as well and is connected to a spark gap (etc.) When the armature is attracted the contact is broken and a high potential difference is set up between the terminals of the secondary.

45.13   Self and mutual inductance

While the changing current in one of two close circuits gives rise to induced currents in the other, it will also give rise to a changing flux, and therefore an induced e.m.f. in itself.

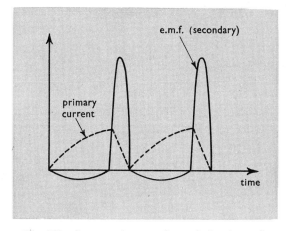

Fig. 268   Input and output for an induction coil

The induced e.m.f. is

$$V = \frac{-\mathrm{d}(N\Phi)}{\mathrm{d}t}$$

where $N\Phi$ is the flux linked with the coil due to the current in itself.

Assuming that the coil is not wound on iron, Φ is proportional to I and $N\Phi \equiv LI$ where L is called the (self) inductance of the coil

$$\therefore \quad V = -L\frac{\mathrm{d}I}{\mathrm{d}t}.$$

Definitions

The practical unit of inductance, the henry, is the inductance of a circuit in which an e.m.f. of 1 volt is induced when the current in it changes at 1 ampere per second.

For a second coil

$$V = -M\frac{\mathrm{d}I}{\mathrm{d}t}$$

where M is the mutual inductance (also measured in henries.)
N.B. the 'order' of the coils does not matter.

The mutual inductance between two circuits is 1 henry when an e.m.f. of 1 volt is induced in one when the current in the other is changing at 1 ampere per second.

46   A.C. CIRCUITS

46.1   The measurement of alternating current and voltage

A moving coil instrument cannot be used with alternating current, as the deflection depends upon the direction of the current.

Hot wire, moving iron or rectifier instruments must be used.

The principle of the first is that the heating effect in a wire (independent of current direction) causes it to expand and this movement can be used to measure the current strength.

In the second, a piece of iron is attracted into a solenoid, or two pieces of iron in a solenoid repel one another, (again independent of current direction).

Both instruments must be calibrated.

The rectifier instrument works as shown in the diagram.

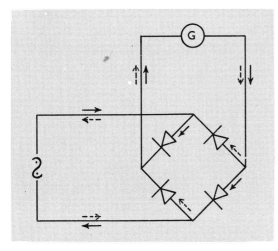

Fig. 269   A rectifier moving coil instrument

However, what value is indicated on an alternating current ammeter, where the current value varies between zero and plus or minus, some maximum value, I_{peak}?

If I^2 is plotted against t, the average value is not zero (as it would be for I).

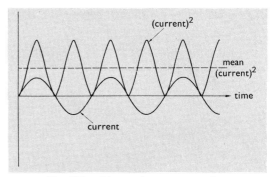

Fig. 270   Alternating current value

Now I^2_{mean} is the value which would be used in calculating the heat losses. Therefore $\sqrt{I^2_{mean}}$, the root-mean-square current value, and a d.c. current of this value would give the same heat losses, so this is the current value indicated by a hot wire (d.c. calibrated) instrument.

It can be shown

$$I^2_{peak} = 2I^2_{mean}$$

so

$$\sqrt{I^2_{mean}} = \frac{1}{\sqrt{2}} I_{peak} = I_{r.m.s.}$$

Alternating current voltmeters likewise are calibrated to read r.m.s. voltage.

46.2   Resistance, capacitance and inductance in a.c. circuits

In each case it is necessary to know the voltage required to send an alternating current through the component. Let the current be $I = I_0 \sin \omega t$. ω is the angular frequency of the alternating current supply, equal to $2\pi f$, where f equals the supply frequency.

46.3   Resistance

Let the instantaneous p.d. across the resistance R, be v.

$$v = RI = RI_0 \sin \omega t$$

$$= V_0 \sin \omega t$$

where $V_0 = RI_0$
and v and I are always in phase.

46.4   Inductance

Let the instantaneous p.d. across the inductance, L, be v.

$$v = L \cdot \frac{\mathrm{d}I}{\mathrm{d}t}$$

$$= L \cdot \frac{\mathrm{d}}{\mathrm{d}t}(I_0 \sin \omega t)$$

$$= L\omega I_0 \cos \omega t$$

Let $V_0 = L\omega I_0$

$$v = V_0 \cos \omega t$$

$$= V_0 \sin \left(\omega t + \frac{\pi}{2} \right)$$

so the voltage leads the current by $\pi/2$, and they are said to be in quadrature. $L\omega$ plays the

part of 'resistance', is measured in ohms, and is called the inductive reactance of the inductance.

46.5   Capacitance

Let the instantaneous p.d. across the capacitor C, be v. Also let the instantaneous charge be q.

$$v = \frac{q}{C} = \frac{1}{C} \int I \cdot \mathrm{d}t$$

$$= \frac{1}{C} \int I_0 \sin \omega t \cdot \mathrm{d}t$$

$$= -\frac{1}{C\omega} I_0 \cos \omega t$$

$$= \frac{1}{C\omega} I_0 \sin \left(\omega t - \frac{\pi}{2} \right)$$

and if $\dfrac{I_0}{C\omega} = V_0$

$$v = V_0 \sin \left(\omega t - \frac{\pi}{2} \right)$$

so voltage and current are again in quadrature, but here current leads the voltage. The 'a.c. resistance' is here $1/C\omega$, the capacitative reactance.

46.6   Resistance, capacitance and inductance in series

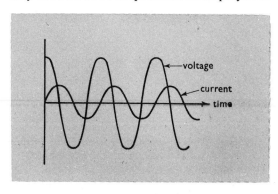

Fig. 271   Voltage and current in an inductor

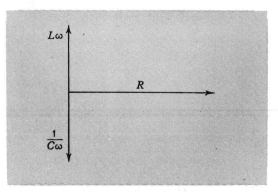

Fig. 272   Resistance, capacity and inductance in series

The total effective 'a.c. resistance' becomes, because of phase

$$\sqrt{R^2 + \left(L\omega - \frac{1}{C\omega}\right)^2}$$

and the whole is called the impedance (the vector sum of resistance and reactance) (symbol, Z)

N.B. $Z = R$ when $L\omega = \dfrac{1}{C\omega}$ $\omega^2 = \dfrac{1}{LC}$

the circuit will then have lowest impedance and is said to be in resonance.

46.7   Some experiments illustrating the properties of capacitors and inductors in d.c. and a.c. circuits

(a)   *Inductor on d.c.*

On closing the switch lamp, L_1 lights appreciably before lamp L_2.

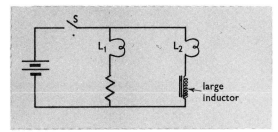

Fig. 273   Delay caused by an inductance in a d.c. circuit

(b)   *Inductor on d.c.*

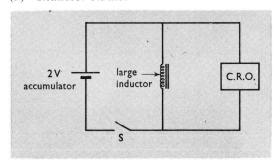

Fig. 274   Voltage pulse created on breaking a d.c. inductive circuit

On opening the switch a large voltage pulse (>300 V) is indicated by the C.R.O. If a plug is pulled from a socket on the inductor to break the circuit (with care!) a large spark is observed.

(c)   *Capacitor on charge*

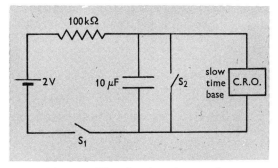

Fig. 275   The charging of a capacitor

On closing switch S_1 an exponential buildup of voltage is observed. The switch S_2 will discharge the capacitor.

(d)   *Capacitor stores charge*

A charged capacitor gives a large flash on discharge. A large electrolytic capacitor charged from a power pack can be used (with care!).

(e)   *Capacitor on d.c.*

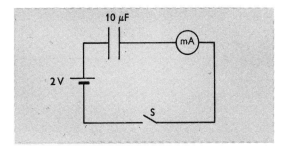

Fig. 276   A capacitor in a d.c. circuit

There is a slight kick on the meter as the capacitor charges, but no continuous current when the switch is closed.

(f)   *Effect of frequency on the reactance of a capacitor*

Working at a fixed voltage an increasing frequency shows an increasing current and therefore a decreasing reactance.

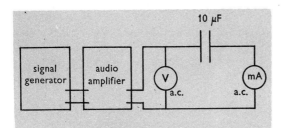

Fig. 277   The effect of frequency on the reactance of a capacitor

Also lowering the capacitance lowers the current.

$$\left(\text{cf. } Z = \frac{1}{2\pi f C}\right)$$

(g) Effect of frequency on the reactance an inductor

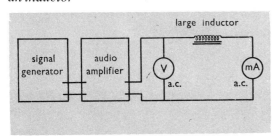

Fig. 278   The effect of frequency on the reactance of an inductor

Working at a fixed voltage, an increasing frequency shows a decreasing current and therefore an increasing impedance.
(cf.

$$Z = 2\pi f L$$

or if R is not negligible

$$Z = \sqrt{4\pi^2 f^2 L^2 + R^2})$$

(h) Lag and lead in an a.c. inductive circuit

Use a long persistence C.R.O. (without trigger) to show that the voltage leads the current. The 2-way, 2-pole switch is worked by hand.

(i) Lag and lead in an a.c. capacitative circuit

A similar circuit to *(h)* is used with a 10 μF capacitor to show that the current leads the voltage.

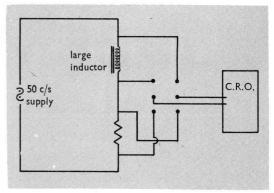

Fig. 279   Lag and lead in an a.c. inductive circuit

(j) Resonance

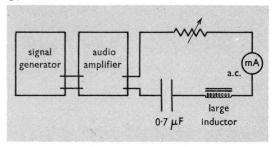

Fig. 280   Resonance circuit

See results on the graph of current against frequency at constant input voltage.

For a 4-V input (measured with a valve voltmeter), there is a potential difference of about 13-V across each component.

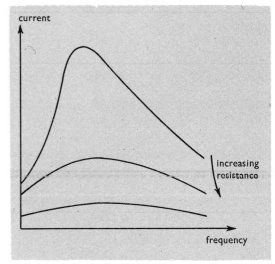

Fig. 281   Resonance curves

MODERN PHYSICS

---

47 CATHODE RAYS

47.1 Electrical discharge through gases

Under normal conditions air is a bad conductor of electricity, otherwise an electroscope would discharge. However, when a very high potential difference is applied (e.g. Van de Graaff machine, lightning) or the distance between electrodes is made very small (mains switches, etc.) then a discharge will occur. In other words a very high potential gradient is necessary to produce an electrical discharge in air at normal pressure.

The discharge under these conditions is a pinkish-purple colour and follows an irregular path of least resistance from one electrode to the other.

If air is placed in a discharge tube and the air pressure gradually reduced then a discharge can be established for a much lower potential gradient.

The features of the discharge continually change with pressure, but a typical appearance, at about 0·3 mm of mercury pressure, is shown in the diagram.

Most of the light comes from the positive column.

As the pressure decreases further the positive column shrinks and the parts of the discharge move to the positive end of the tube until finally Crooke's dark space 'fills' the tube and there is little to be seen.

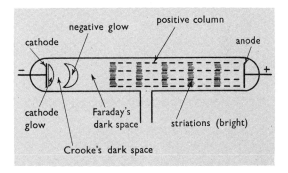

Fig. 282 Electrical discharge through air at low pressures

However, under these conditions an important new effect is noticed, the glass round the cathode fluoresces (usually green, but this depends on the chemical composition of the glass). Moreover the nature of this discharge is independent of the original gas in the discharge tube.

This fluorescence was thought to be due to the emission of 'something' from the cathode, and originally this something was called 'cathode rays'.

As cathode rays are not easily produced in this manner another way of producing them will be considered first, using a much smaller potential gradient, and heating the cathode to assist their release.

47.2 Thermionic emission

It is found that if the cathode is heated (by passing an electric current through it), then 'cathode rays' are produced for a smaller potential gradient across the tube. Thus cathode rays are produced more easily in 'hot-cathode' tubes and these can be used to demonstrate the properties of cathode rays.

47.3 The properties of cathode rays

(*i*) They cause fluorescence when they fall on appropriate materials.

(*ii*) They travel in straight lines (Maltese-cross experiment).

(*iii*) They can be deflected by magnetic or electrostatic fields (Deflection tube).

(*iv*) When they strike an object they cause (*i*) a rise in temperature, (*ii*) a mechanical force.
(Paddle-wheel experiment).

(*v*) They affect photographic plates.

(*vi*) They carry negative charge (Perrin tube experiment)

47.4 The nature of cathode rays

All of these properties are most easily explained by the idea that cathode rays are a stream of negatively charged particles. The greatest support for this comes, however, from the fact that the ratio of charge to mass (*e/m*) can be measured and shown to be constant.

Many methods of measuring *e/m* are now available with varying degrees of accuracy and complexity. One which is nearest to the rather more complicated original of J. J. Thomson's (1897) will be considered.

The hot-cathode deflection tube is used and first the deflection is found when only an electrostatic field is applied between the plates EE. Then a magnetic field is applied by a pair of Helmholtz coils to bring the cathode ray track back to zero deflection.

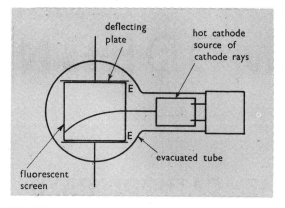

Fig. 283 The measurement of *e/m* for cathode rays

If the applied p.d. is *V*(volts) and the plates are *x* (metres) apart, the field is

$$E = \frac{V}{x}$$

If a particle of charge *e* (coulombs) and mass *m* (kg) traverses the field of length *l* (metres), in time *t* (s), at velocity *v* (horizontal), then

$$t = \frac{l}{v}$$

and the vertical acceleration is

$$\frac{eE}{m} = a$$

then the deflection is

$$\tfrac{1}{2}at^2 = \tfrac{1}{2}\frac{eE}{m}\left(\frac{l}{v}\right)^2$$

so

$$\text{deflection} = \frac{eEl^2}{2mv^2}$$

For the second part
Force on moving charge = $Bev = Ee$

so

$$v = \frac{E}{B}$$

where *B* = the magnetic flux density produced by the coils

$$B = \mu_0 \frac{8NI}{5\sqrt{5}r}$$

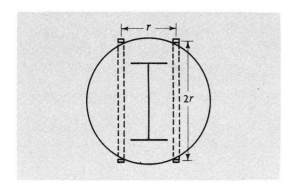

Fig. 284　Helmholtz coils for the e/m experiment

where $\mu_0 = 4\pi \times 10^{-7}$
$\qquad I$ = current in amps
$\qquad N$ = number of turns on one coil

thus v may be determined and by substituting into the first expression, e/m may be calculated.

Thomson's original experiment and others which followed it supported the idea that cathode rays were particulate in nature and that e/m was independent of the materials of the tube. This also suggests that these particles were fundamental constituents of matter, which are now called electrons.

47.5　The charge on the electron

The classic experiment was performed by Millikan who investigated the motion of a charged drop in an electrostatic field.

Two parallel metal plates were mounted so that oil drops could be sprayed in through

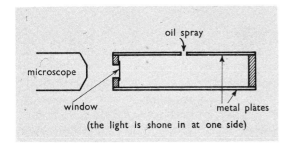

Fig. 285　Millikan's apparatus

a hole in the top plate. An electrostatic field is set up between them. The motion of a single oil drop is followed in a microscope using side illumination.

First the time for the drop to fall freely between two marks on the microscope scale is noted. Then a suitable field is applied and the time for the drop to rise between the same marks is also noted. The same drop is followed over a long period of time, the period of each traverse being recorded.

While the drop is falling freely it travels with a constant terminal velocity due to the viscosity of the air.

Quantitatively, therefore, making use of Stoke's law that

$$F = 6\pi r\eta v$$

and when the terminal velocity is reached,

F = apparent weight of drop

$$= \tfrac{4}{3}\pi r^3(\rho - \sigma)g$$

$$\therefore \quad 6\pi r\eta v = \tfrac{4}{3}\pi r^3(\rho - \sigma)g$$

where r = radius of drop
$\qquad \eta$ = viscosity of air
$\qquad v$ = velocity
$\qquad F$ = force on drop
$\qquad \rho$ = density of oil
$\qquad \sigma$ = density of air

whence the radius, r, may be found.

When the field, E, is applied

$$E = \frac{V}{d}$$

where V = applied p.d.
$\qquad d$ = separation of plates

and upward force = Ee − apparent weight of drop

$$\therefore \quad Ee - \tfrac{4}{3}\pi r^3(\rho - \sigma)g = 6\pi r\eta v'$$

where v' = upward terminal velocity. Whence e may be found.

The results of this experiment always give e as a multiple of about 1.60×10^{-19} coulomb. If during a single run when many observations are taken the charge on the drop changes due to absorption of ions, etc. it always changes by this amount or a multiple of it. This experiment strongly supports, therefore, that there is a fundamental unit of electric charge which is identified with the charge on the electron.

47.6   The relations with electrolysis

The faraday constant has already been defined as the charge necessary to release one mole of a substance (e.g. 1 g of hydrogen).

Also Avogadro's number, N_A, is defined as the number of atoms in 12 g of carbon $-$ 12 (or more loosely as the number of atoms in one mole of the substance.)

Now if the idea that ions are formed in electrolysis by the loss or gain of electrons is used, these two can be connected. For if e is the electronic charge, and each hydrogen ion has one charge

$$N_A \cdot e = F$$

hence a further check is possible by comparison with Millikan's experiment. Both 'e's' are found to be the same, linking the electron with the unit ionic charge.

48   THERMIONIC VALVES

48.1   Thermionic emission

The emission of electrons from a heated wire is due to them acquiring sufficient energy to escape. The emission is increased if the temperature is raised, or if the wire is coated with suitable materials.

The emitted electrons repel further electrons which are trying to escape and the wire becomes positively charged, also attracting them back. This swarm of electrons held near the wire is referred to as the space charge. Only if a high positive potential is applied to a plate or anode near the wire will the electrons flow to the anode. Such an arrangement is known as a diode.

48.2   The diode

The behaviour of a diode can be investigated with the circuit shown.

Both anode voltage and filament current can be varied, and the anode current measured. However it is usual to work with a fixed filament current, so that the variation of anode current with anode voltage is investigated.

The filament current is higher in curve 2 than in curve 1.

The maximum or saturation current is only reached for a high anode voltage when all the emitted electrons are reaching the anode. Both the current and the voltage required are higher for a hotter filament. In the earlier part of the graph $I_A \propto V_A^{3/2}$ (three-halves power law).

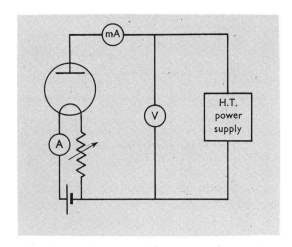

Fig. 286   Circuit to investigate a diode

48.3   The diode as a rectifier

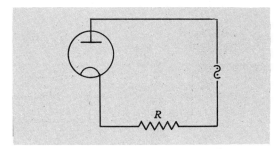

Fig. 288   The diode as a rectifier

If an alternating e.m.f. is applied, current will only flow through the resistance when the anode is positive relative to the filament. The current will be intermittent (See Fig. 289 below), and a smoothing circuit is needed to give a reasonable direct current:

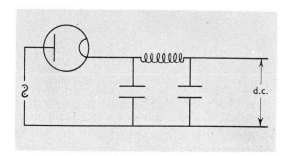

Fig. 290   A half wave rectification plus smoothing circuit

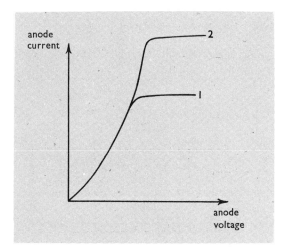

Fig. 287   Diode characteristics

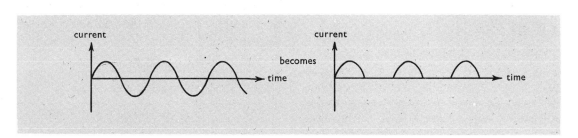

Fig. 289   Half wave rectification

A double-diode can be used to give full-wave rectification

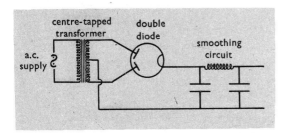

Fig. 291   A full wave rectification plus smoothing circuit

The wave form becomes:

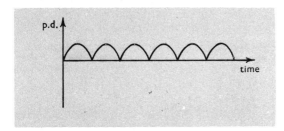

Fig. 292   Full wave rectification

before smoothing.

The diode can also be used as a detector/rectifier in a radio set. The aerial signal is high frequency alternating current with the amplitude varying at audio frequencies, and the response of a pair of headphones is zero, as in one cycle of the carrier wave the plate of the headphones does not have time to move. However, if it is rectified there will be a non-zero response, as the headphones integrate the unidirectional pulses.

48.4   The triode

A third electrode was added to the diode to make a triode. It took the form of an open mesh of wire between the filament or cathode and the anode. Small potentials applied to this grid have a large effect on the electron flow, for, when the grid is made positive the effect of the space charge is reduced and vice-versa.

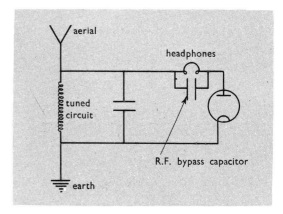

Fig. 293   The diode as a detector of radio signals

The filament current of a triode is usually kept constant.

48.5   Triode characteristics

The circuit used to investigate the behaviour of the valve is shown.

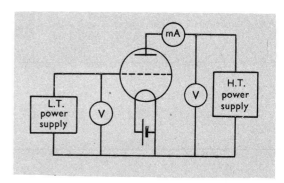

Fig. 294   Circuit to investigate a triode

Various graphs can be drawn, e.g. anode current (I_a) against grid voltage (V_g) for fixed anode voltages (V_a), or anode current against anode voltage for fixed grid voltages. Such graphs are called the characteristics of the valve.

Valve constants are defined as follows:

(*i*)   Anode slope resistance

$$= \frac{\delta V_a}{\delta I_a} = \rho = \frac{\text{a small change in anode voltage}}{\text{the resulting change in anode current}}$$

for a constant grid voltage.

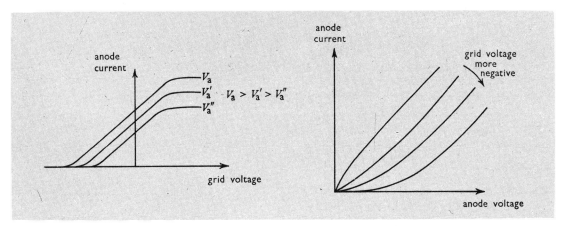

Fig. 295   Mutual characteristics of a triode   Fig. 296   Anode characteristics of a triode

(*ii*)   Amplification factor

$$= \frac{\delta V_a}{\delta V_g} = \mu$$

$$= \frac{\text{the change in anode voltage needed to produce a certain change in anode current}}{\text{the change in grid voltage needed to produce the same change in anode current}}$$

(*iii*)   The mutual conductance

$$= \frac{\delta I_a}{\delta V_g} = K$$

$$= \frac{\text{the change in anode current}}{\text{the change in grid voltage necessary to cause this change}}$$

and

$$\mu = K\rho$$

These constants are stated strictly for one set of conditions only, but as the characteristic curves are practically parallel over a wide region, they may be taken as constants over this region.

48.6   A simple expanation of the triode characteristics

With large negative grid voltage the effective space charge is sufficiently high to repel all the emitted electrons and prevent any current flowing through the valve. The space charge falls as the grid voltage increases through to zero and so the value of anode current increases. Finally saturation is reached when all the emitted electrons reach the anode.

48.7   The triode as an amplifier

If a triode is to be used as an amplifier a large load resistance is connected in the anode circuit.

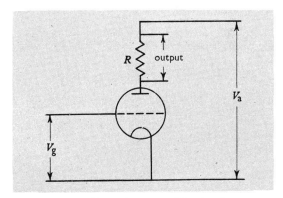

Fig. 297   The voltage amplification factor

A change in grid voltage causes a change in anode current, and therefore a change in voltage across the load resistance which is used. However, this also means a change in anode voltage (keeping the supply voltage constant) so if, for example, the grid voltage goes less negative, the anode current increases, the potential difference across the load resistance increases, so the potential difference across the valve decreases, so lessening the anode current. Thus the amplification is less than in the static case.

When an a.c. input is applied, only a.c. changes from the load resistance are used, and the d.c. voltage can be ignored in calculations.

The a.c. change δV_g has the same effect as a change $\mu\delta V_g$ in anode voltage. This is effectively in the circuit valve, load, and power supply, and causes an a.c. current

$$\frac{\mu\delta V_g}{\rho + R}$$

and an a.c. voltage across R of

$$\frac{\mu R\delta V_g}{\rho + R} = \delta V_R$$

and the (a.c.) voltage amplification factor

$$\frac{\delta V_R}{\delta V_g} = \frac{\mu R}{\rho + R}$$

If R gets larger this increases, but a higher value of V is needed.

For maximum power

$$\rho \simeq R.$$

48.8   Grid bias

In deriving the above relationships, the straight part of the characteristics has been used. So the a.c. variations must not take the grid voltage off this part. So the grid is biased to a negative value, about which a.c. changes are applied.

This used to be done by a battery, but now a resistance in the cathode to negative line is used, which makes the cathode slightly positive relative to the negative line, and therefore to the grid.

48.9   The cathode-ray oscilloscope

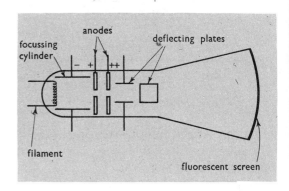

Fig. 298   The cathode ray tube

This is a development of J. J. Thomson's apparatus. Electrons are emitted by the hot filament and accelerated by the two anodes. These, with the focussing cylinder, also help to focus the beam. This part of the tube is referred to as an electron gun.

The electron beam can be deflected by voltages applied to the mutually perpendicular plates, or by currents flowing in coils placed outside the tube. The plates causing the vertical deflection are called the Y-plates, and those which cause the horizontal deflection are the X-plates.

Often a time base is connected to the X-plates. This is an electronic arrangement whereby (if there is no signal on the Y-plates) the spot on the fluorescent screen moves steadily across the screen from left to right and then rapidly flies back to the left again. The frequency is usually adjustable, so that it can be matched to any signal on the Y plates and thereby produce a steady trace. This makes a very versatile instrument widely used in physics.

49 THE PHOTO-ELECTRIC EFFECT. X-RAYS

49.1 The photo-electric effect

A freshly cleaned and amalgamated zinc plate is connected to an electroscope. If it is negatively charged and illuminated with ultraviolet light it loses its charge, but if it is positively charged there is no loss of charge.

The u.v. light must fall on the plate, not on the air near it. This suggests a surface effect.

This effect is due to the emission of electrons which are in excess on the negatively charged plate.

The particles can be proved to be electrons by the measurement of e/m by advanced methods.

If a glass plate is placed in front of the u.v. lamp, cutting off the u.v. part of the radiation, no loss of charge can be detected. So visible light of a lower frequency does not cause photoelectric emission in this case. In general it is found that there is always a minimum frequency for the photoelectric effect to start, and once this frequency is exceeded photoelectrons are emitted as soon as the radiation falls on the surface.

These last two points are in direct contrast with a classical explanation. This would be in terms of the atoms gradually acquiring energy from any incident wave until there was sufficient to emit the electron. Calculation shows that this would mean a long time interval whilst the energy built up, far longer than any observed.

Thus classical physics breaks down and the correct explanation of the photoelectric effect is due to Einstein's application of Planck's hypothesis that light radiation is quantized. Light travels as a large number of quanta of energy or photons, each carrying energy $h\nu$ (h = Planck's constant and ν = frequency).

When the light falls on a surface the quantum of energy is given to a single electron which may be emitted if the photon carries sufficient energy.

Einstein produced the equation

$$(\tfrac{1}{2}mv^2)_{max} = h\nu - W$$

$(\tfrac{1}{2}mv^2)_{max}$ is the maximum kinetic energy of the photoelectron, and W the work function of the surface, i.e. the energy needed to remove the electron from the metal. Usually a smaller energy is observed due to collisions during the escape process.

If $(\tfrac{1}{2}mv^2)_{max} = 0$, $h\nu = W$ and this gives a minimum or threshold frequency, as found experimentally. Thus the quantum theory fits the experimental facts. (Einstein won the 1921 Nobel prize for physics for this work).

This phenomenon is applied in the photoelectric cell and in the photovoltaic cell.

49.2 Further photoelectric experiments

Further experiments on this law can be made as follows:

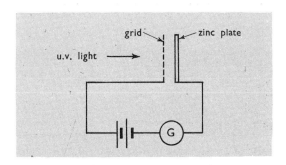

Fig. 299 Experiments on the photoelectric effect

A grid is placed near to the zinc plate and a sensitive galvanometer used to measure the current, due to photoelectrons. If the grid is made increasingly negative the current is steadily reduced as fewer and fewer electrons have sufficient energy to overcome the potential barrier.

Einstein's equation can be written

$$h\nu - W = eV_0$$

where V_0 is the maximum negative potential that can be overcome.

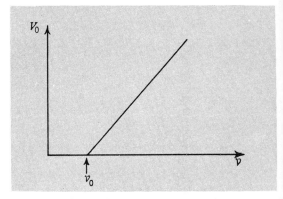

Fig. 301   The variation of 'cut-off' potential with frequency

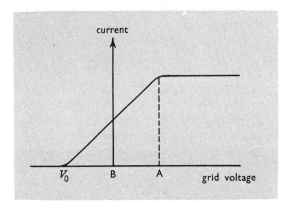

Fig. 300   The effect of grid voltage on the current due to photoelectrons

N.B. The true zero of V is at A, not B, the difference being due to contact potential.

It is found that V_0 depends only ν, whereas the saturation current depends on the intensity of the light.

Millikan measured V_0 as a function of ν, carefully correcting for contact potential, as

$$V_0 = \frac{h\nu}{e} - \frac{W}{e}$$

the gradient gives h and the intercept W (if e is known) and results are consistent.

49.3   X-rays

In 1895 Röntgen, while experimenting with a discharge tube, found effects, such as fluorescence and the fogging of photographic

plates, occurred at distances too far from the tube to be explained by cathode rays. He called the responsible radiation X-rays and went on to investigate the properties of this radiation.

49.4   The properties of X-rays

(*i*)   X-rays are always produced when cathode rays (electrons) strike any solid body.

(*ii*)   X-rays cannot be deflected by magnetic or electrostatic fields.

(*iii*)   X-rays travel in straight lines.

(*iv*)   X-rays will penetrate most materials but are gradually absorbed especially by dense materials.

(*v*)   X-rays cause fluorescence and effect photographic plates.

(*vi*)   X-rays cause ionization and therefore discharge positively or negatively charged objects.

49.5   The production of X-rays

The properties were found in cold-cathode tubes, which have now been replaced by hot-cathode types.

The Coolidge type of X-ray tube is highly evacuated and electrons are emitted from a heated cathode, accelerated by a very high

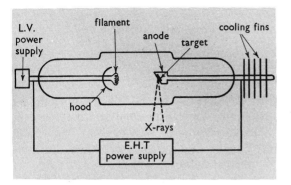

Fig. 302   An X-ray tube

potential difference (tens of thousands of volts) and fall on the anode. A hood assists in focussing. The X-rays are emitted through a window in the side of the tube.

The anode is a copper block containing a tungsten block or target set at 45° to the electron beam and then the X-rays, which are mainly produced at right angles to the electron beam, come out through the window. A large amount of heat is produced but the tungsten is of high melting point and the copper rod a good conductor. Heat is dissipated from the cooling fins, assisted in high intensity tubes by oil circulating near the target area and by having a rotating anode so that one part of the target is not continuously heated.

49.6   The uses of X-rays

(a)   Medical

X-rays are widely used for investigation and treatment.

(b)   Industrial

X-rays can assist in inspection of products or welds, etc.

(c)   Research

Among many techniques, that of X-ray diffraction is very important.

49.7   X-ray diffraction

It was thought that the natural spacing of atoms in a crystal might be about the same as the wavelength of the X-rays, and so the crystal should behave like a diffraction grating. This was observed about 1913 by Friedrich, Knipping and von Laue.

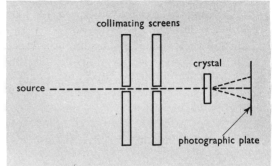

Fig. 303   X-ray diffraction

A narrow beam of X-rays, selected by pinholes in two lead screens, fell on a crystal. The resulting exposure showed not only the expected centre spot but a series of further spots in a regular pattern around it.

This supports the belief that X-rays were electromagnetic waves and that crystals are regular three-dimensional arrays. The work won von Laue (the group leader) a Nobel prize in 1914.

The connection between diffraction and crystal structure was first studied by Sir William Bragg and his son, and their work led to Bragg's law.

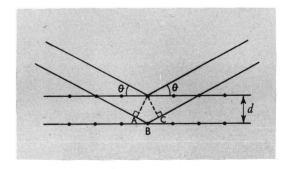

Fig. 304   Bragg's law

The crystal is taken as being a regular array of atoms, forming a series of planes in

M

which again the atoms are regularly spaced.

X-rays falling on to the atoms are scattered and only when angles of incidence and scattering are equal does reinforcement occur for atoms in a given plane. This is the same situation that occurs in the reflection of light according to Huygen's construction.

When several atomic planes are concerned as shown, X-rays scattered from different layers will interfere and reinforcement occurs when the path difference equals a whole number of wavelengths.

$$\text{path difference} = AB + BC$$
$$= 2d \sin \theta$$
$$= n\lambda$$

and

$$2d \sin \theta = n\lambda \quad \text{(Bragg's law)}.$$

where d = separation of planes

θ = glancing angle

Bragg tested this with his spectrometer.

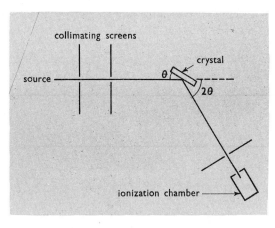

Fig. 305   The verification of Bragg's law

X-rays are collimated by two slits (in lead plates) and fall on the crystal. An ionization chamber is placed at an angle with the incident beam equal to twice the glancing angle. The ionization current produced is a measure of the X-ray intensity and it is found that maxima occur when $\sin \theta_1 : \sin \theta_2 : \sin \theta_3$ are as $1 : 2 : 3$, thus verifying Bragg's law.

49.8   X-ray spectra

Using the principle of Bragg's spectrometer the wavelengths of the X-rays present and their intensity can be analysed if the crystal spacing is known (or vice versa). The results of Bragg's and others' work can be summarized as follows:

An X-ray tube produces two types of radiation:

(*i*)   A line spectrum characteristic of the target material.

(*ii*)   A continuous spectrum which does not depend on the target material, but does depend on the tube voltage.

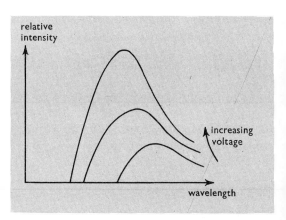

Fig. 306   Continuous X-ray spectra

With the continuous spectrum there is a minimum wavelength of emitted X-rays.

This is again a quantum phenomenon. If it is assumed that all the kinetic energy of the electron is converted into the energy of the X-ray photon, then:

$$h\nu = \tfrac{1}{2}mv^2$$

where h = Planck's constant
ν = X-ray frequency
m = mass of electron
v = velocity

but in travelling between electrodes at a potential V apart an electron gains energy eV (from the definition of potential difference)

so $$\tfrac{1}{2}mv^2 = eV = h\nu$$

and if all this kinetic energy is given to an X-ray photon, it will have maximum energy and frequency and therefore minimum wavelength.

$$\therefore \quad h\nu_{max} = eV$$

and so ν_{max} or λ_{min} depends on V

$$\lambda_{min} = \frac{c}{\nu_{max}} = \frac{ch}{eV}$$

where $\nu_{max} = $ maximum X-ray frequency
$\lambda_{min} = $ corresponding minimum X-ray wavelength

(N.B. The higher the frequency the more penetrating the radiation.)

50   RADIOACTIVITY

50.1   The discovery of radioactivity

In 1896 Henri Becquerel was investigating the fluorescence of uranium to find if X-rays were emitted at the same time. However, he found that the greatest effect on a photographic plate occurred when some uranium salt had been left near it, but in the dark so that no fluorescence, etc. was possible.

Further study showed this phenomenon to be unconnected with fluorescence and a new phenomenon had been discovered, later called radioactivity by Marie Curie.

Investigation of this new phenomenon, particularly by studies of penetration and the effect of magnetic fields, showed that there were three distinct types of radioactivity.

Type 1   α-rays, were the least penetrating, being stopped by a sheet of paper or a few centimetres of air

Type 2   β-rays, were more penetrating, but were stopped by a few millimetres of aluminium or about a metre of air.

Type 3   γ-rays, were very much more penetrating, and required several centimetres of lead to stop most of them.

A hypothetical experiment shows the effect of a magnetic field.

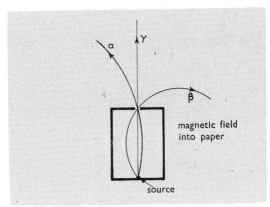

Fig. 307   Deflection of radioactive radiations in a magnetic field

The α-rays are slightly deflected, and because of the direction of bending must be positively charged.

The β-rays are greatly deflected, and must be negatively charged.

The γ-rays are undeflected.

50.2   The detection of radioactivity

Various methods may be applied to the detection of radioactivity.

(a)   *Scintillation methods*

   (i)   *Spinthariscope*

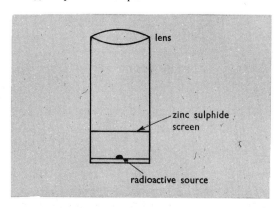

Fig. 308   The spinthariscope

The eye, when adjusted to dark conditions, can see flashes of light over the zinc sulphide screen. These are due to bombardment by α-rays from the source, causing the emission of light.

The random nature of the flashing is the first indication of the random nature of radioactive decay.

Measurements can be made for weak sources, but the procedure is very tiring.

   (ii)   *The Scintillation Counter.* The principle is shown by the block diagram.

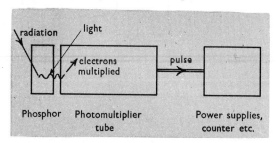

Fig. 309   The principle of the scintillation counter

Radiation falling on the phosphor causes the emission of light which strikes the photocathode of the photomultiplier tube. This causes the photoelectric emission of an electron. The photomultiplier tube increases this electron into a shower of electrons or an electric pulse which can be counted electronically.

(b)   *Ionization methods*

   (i)   The Ionization Chamber. When radiations pass through a gas, collisions will occur and electrons are removed from many atoms to leave positively charged ions. If an electric field is applied these ions will be collected and a small ionization current can be detected.

As the current is very small a suitable detector is the Wulf electroscope.

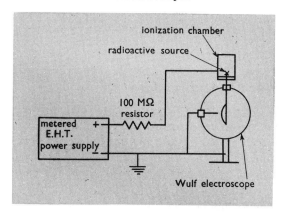

Fig. 310   The investigation of the behaviour of an ionization chamber

The potential difference across the ioniza-chamber is varied and the number of discharges per unit time is a measure of the ionization current flowing.

In region A some ions produced recombine and therefore an increasing field, resulting in quicker ion movement, causes an increasing current.

In region B, all the ions reach the electrodes and saturation has occurred. The ionization chamber would usually be operated in this region. In region C the ions are caused to move so quickly that more ions are produced

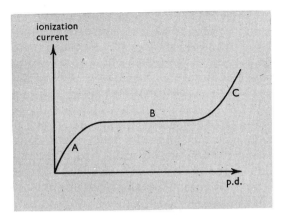

Fig. 311   The characteristics of an ionization chamber

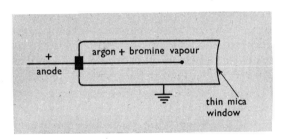

Fig. 312   A Geiger Müller tube

by collision and 'gas amplification' is said to occur.

(*ii*)   The Geiger Müller Counter. This is a development of the ionization chamber and depends on gas amplification.

A central wire is raised to a high positive potential and the case is earthed. A thin window permits the entry of the radiation. The ions produced are amplified by the process described above and a current pulse is produced which can be counted electronically.

In region D, 'the proportional region' amplification occurs but the output pulse still depends on the energy of the entering particle.

In region E 'the Geiger region' the output pulse is independent of the energy of the entering particle. The tube is normally worked in the centre of this plateau.

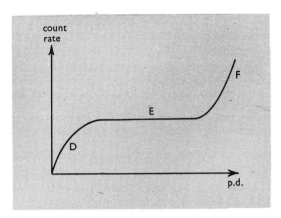

Fig. 313   The characteristics of a Geiger Müller counter

In region F a continuous discharge can occur which may destroy the tube.

The trace of bromine acts as a quenching agent, stopping the discharge after the first main pulse and preventing multiple discharges. The 'halogen' type of tube works at lower voltages than the 'organic' type.

(*iii*)   The Cloud Chamber. This depends on the fact that a supersaturated vapour will condense on ions and form a track.

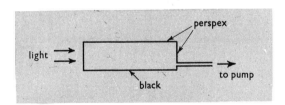

Fig. 314   The principle of the cloud chamber

A little alcohol and water is placed inside the cloud chamber and on pulling out a hand vacuum pump, the adiabatic expansion produces cooling and supersaturation.

If radiation has passed through the chamber just before expansion then the condensation occurs on the ion tracks. A potential difference of about 200 volts is applied between a ring round the top and the base to clear ions after expansion.

(c) Photographic methods

Special photographic plates with thicker and denser emulsions than usual are used. The tracks are very short and need to be studied microscopically.

50.3 The identification of the three types of radiation

Using a radium source, a thin window Geiger tube and a scaler or ratemeter, it is possible to investigate the effect of placing absorbers in the path of the radiation. It should be possible to show two discontinuities in the graph of count rate against thickness. (Thickness is measured in mg cm^{-2}.) This indicates three types of radiation.

Using separate sources the effect of a magnetic field can be investigated.

(a) α-rays

So far it has been found that these are weakly penetrating particles of positive charge and causing great ionization.

The final identification depended on Rutherford and Royds' experiment. The principle is as follows:

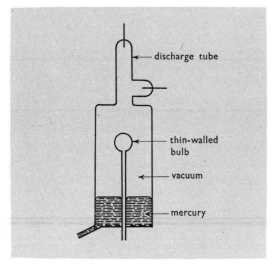

Fig. 315 Rutherford and Royds' experiment

A small quantity of radon (an α-ray emitting gas) was enclosed in a thin-walled bulb. The α-rays passed through the walls of this bulb into a surrounding vacuum. After several days the mercury was raised, compressing the gas, which was found to have accumulated, into the discharge tube. This showed the characteristic spectrum of helium.

A further test was made with helium in the bulb to prove that no diffusion had occurred.

Thus α-rays must be positively charged helium. Measurements of the charge to mass ratio were made by magnetic and electrostatic deflection. This showed half the value for a hydrogen ion, so the α-ray must be a helium ion of double charge, and this will be shown later to be equivalent to a helium nucleus. Hence 'α-ray' should be α-particle.

(b) β-rays

If the β-radiation from a well collimated source is deflected by a magnetic field the resultant beam is found to be spread out. This is because β-rays are emitted with a spread of velocities and as the deflection depends on this velocity the beam must spread. More elaborate experiments using parallel electrostatic and magnetic fields (compare notes on positive rays) enable both velocity and ratio of charge to mass, or specific charge to be found.

The results show the same value for specific charge as for electrons and the velocity to be very large (a few tenths of the velocity of light). Thus β-rays are high speed electrons.

(c) γ-rays

By comparison with X-rays, γ-rays are found to be similar but even more penetrating, supporting the idea that γ-rays are electromagnetic waves.

50.4 The reactions of radiations with matter

(a) α-particles

These have small penetrating power, but any one type of α-particle (i.e. emitted in a particular case) has a definite range which is usually a few centimetres of air and can be measured in various ways.

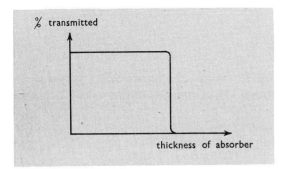

Fig. 316 The absorption of α-particles

One typical experiment uses the ionization chamber in a form such that the height can be altered.

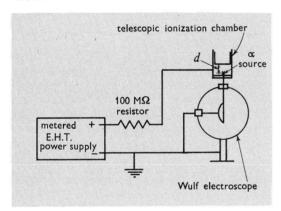

Fig. 317 Investigation of the range of α-particles

The distance from the source to the lid of the chamber is altered and the pulse rate noted. Until this distance is greater than the range the pulse rate should increase as more ionization can occur. Graphical treatment of the results gives the value of the range.

(b) β-particles

These do not have such a well defined range. If the intensity of radiation transmitted through a series of absorbers is measured with a Geiger counter a gradual falling off is found, but there is a maximum range.

B = Bremsstrahlung (braking radiation), X-rays produced in the absorber as the β-particles are slowed down.

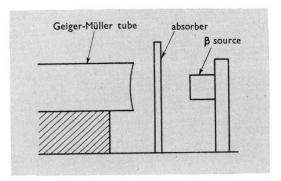

Fig. 318 To investigate the absorption of β-particles

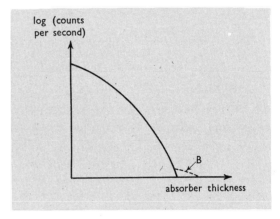

Fig. 319 Effect of absorbers on β-particle transmission

N.B. If the thickness is given in mg cm⁻², the material used is of little importance.

(c) γ-rays

The process of interaction is quite different as energy is usually given up in a few large

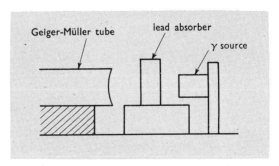

Fig. 320 To investigate the absorption of γ-rays

amounts rather than in many small ones. The γ-rays therefore carries on unchanged or is drastically altered. This makes the whole

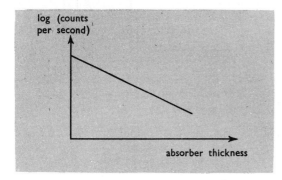

Fig. 321 Effect of absorbers on γ-ray transmission

process much more 'chancy' and there is no simple range.

50.5 The biological effects of radioactivity and safety

When radiations from radioactive materials pass through a living cell they can cause damage. This damage may be either to the material of the cell, which, provided the damage is not too great at any one time can be repaired by the cell, or to the genetic mechanism which cannot be repaired.

There is no safe dose for the human cell and all unnecessary exposure should be avoided. However, it is impossible to eliminate all exposure because of the background radiation to which everyone is subject. This radiation is due to cosmic rays, fall-out from atomic weapons, etc., and radioactive material in the earth.

α-particles can do little harm if the source is outside the body, as they are stopped by the outer skin, but they can do a great deal of harm if they are absorbed into the body.

β-particles and γ-rays can penetrate to varying extents.

Radioactive sources are divided into sealed (such as used in physics where the radioactive material is not accessible) and unsealed or open sources (similar to ordinary chemicals).

Precautions relating to sealed sources are:

(*i*)   No eating or drinking in the laboratory.

(*ii*)   Wear a laboratory coat at all times.

(*iii*)   Never handle a radioactive source in such a way that the source comes into contact with the skin, and in no circumstances should radioactive sources be held close to the eye for examination.

(*iv*)   Keep all sources, when not in use, in their containers in a locked place.

(*v*)   All containers of radioactive materials including sealed sources must be clearly labelled.

These rules apply whatever the strength of the source involved.

50.6 Statistics and decay

All radioactive experiments suggest that the radioactive decay process is a random one.

If a special experiment is done then the results can be statistically analysed.

For this a suitable source is placed in front of a Geiger Müller tube connected to a scaler and the counts recorded in at least twenty separate minute intervals.

The results may be analysed as follows:

(*i*)   Find the average count rate per minute, $\bar{x}$

(*ii*)   For each separate measurement calculate the deviation $x - \bar{x}$

(*iii*)   Square these deviations, add the squares, divide by the number of readings less one and take the square root.

This gives the standard deviation,

$$\sigma = \left[\frac{\sum_{1}^{n} (x - \bar{x})^2}{n-1}\right]^{1/2}$$

Observations

(*i*)   Statistical theory suggests that $\sigma \simeq \sqrt{\bar{x}}$. This should be checked from the results obtained.

(*ii*) Theory suggests that 31·7% of the observations will be different from $\bar{x}$ by more than σ and 50% will be different from $\bar{x}$ by more than the probable error ($= 0.674\,\sigma$)

Again a comparison can be made.

Agreement between theory and practical results will support the idea of radioactive decay as a truly random process.

The other main point is that the larger the count the more reliable it is ($\sigma \simeq \sqrt{\bar{x}}$) but that in any particular result there is little point in quoting too many figures.

50.7   Decay and half-life

Radioactivity results from atoms breaking down and going to a more stable state. This is a random process as one atom has no effect on another. However, the process can be treated statistically.

Suppose that a given sample of radioactive material contains N radioactive atoms and that the probability that each atom will decay in any one second is a constant (the decay constant) λ. Then the average number dN that will decay in a small time interval dt is given by

$$dN = -\lambda N \cdot dt$$

Integrate from $t = 0$ to $t = t$

where N_0 = number of atoms at time 0.

N_t = number of atoms left after time t

$$\int_{N_0}^{N_t} \frac{dN}{N} = -\lambda \int_0^t dt$$

$$N_t = N_0 e^{-\lambda t}$$

The decay constant is equal to the fraction of the total number of atoms which decay in unit time, but is not much used in practice.

An alternative form, the half-life ($t_{1/2}$) is used, defined as the time in which the amount of radioactive material decays to one half of its original value

i.e.

$$N_t = \frac{N_0}{2}$$

$$\therefore \quad \frac{N_0}{2} = N_0 \exp(-\lambda t_{1/2})$$

$$\therefore \quad \tfrac{1}{2} = \exp(-\lambda t_{1/2})$$

$$\therefore \quad \log_e\tfrac{1}{2} = -\lambda t_{1/2}$$

$$t_{1/2} = \frac{\log_e 2}{\lambda} = \frac{0.693}{\lambda}$$

The half-life is characteristic of a particular radioactive material and can vary over a tremendous range from millions of years to a millionth of a second or less.

50.8   The determination of half-life

Thoron, with a convenient half-life of about 54 seconds is often used as an example. It is an α-particle emitting gas produced from thorium compounds by a previous radioactive process.

Various procedures are possible, the one described is very simple.

Thoron is produced by keeping some thorium hydroxide in a bag in a polythene squeeze bottle and when needed the bottle is pressed and some thoron expelled. At other times the bottle should be kept clipped up.

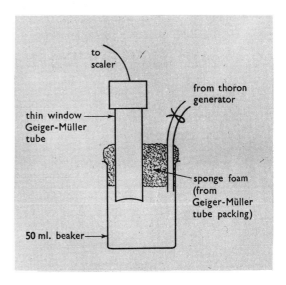

Fig. 322   To investigate the half-life of thoron

The apparatus is set up as shown and some thoron sent into the beaker. The scaler is read (to the nearest 10) every 10 seconds and a graph of counts against time plotted.

Then the half-life can be read off in several places from the graph as shown.

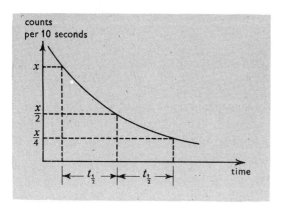

Fig. 323 Decay of thoron

Alternately the results can be treated by plotting log (counts per ten seconds) against time.

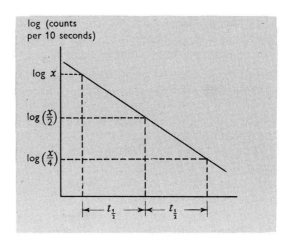

Fig. 324 Decay of thoron (log, plot)

50.9 The strength of radioactive sources

This is measured in curies. If any piece of radioactive material undergoes $3 \cdot 7 \times 10^{10}$ disintegrations per second it has an activity of one curie. This is a large unit and all school sources must be less than 10 micro-curies.

51 THE STRUCTURE OF THE ATOM

51.1 The scattering of α-particles

Geiger and Marsden (under Rutherford) found that when a pencil of α-particles fell on to a thin metal foil, the great majority passed through with little deflection but some were scattered appreciably whilst a few (1 in 8000) were deflected back along their path.

These results were in great contradiction to the then accepted theory (the 'plum-pudding' model).

Rutherford put forward the idea of the 'nuclear' atom where the atom is thought of as containing a central nucleus, of radius about 10^{-14} m, surrounded by electrons forming an atom of radius about 10^{-10} m. When an α-particle reaches the foil it will most probably take a path far from all the nuclei and be unaffected. There is a chance of it going near to a nucleus and being deflected, and an even smaller chance of a near head-on collision giving the few deflected back particles which Geiger and Marsden observed.

Rutherford developed the theory more mathematically, and Geiger and Marsden then tested it and found that the results fitted.

The Rutherford formula also enables a value for the central charge to be found as a multiple of a basic unit. Difficult experimental work, led to values which were not half the atomic weight as then expected; however, they did fit with results from Moseley's work on X-rays.

51.2   Characteristic X-Rays and Moseley's law

X-ray spectra, as was mentioned above, consists of two parts, and it is the line spectra which is now to be considered. Moseley found that the lines for different target elements always appeared in similar groups, a short wavelength group, the K series of two lines, and then a longer wavelength group, the L series of three lines. Moseley concentrated on the first line of the first group, K_α.

Using 38 elements he first found that there was a rough connection between the square root of the frequency of the K_α line and the atomic weight of the elements, but the resulting graph was not a good straight line, there was too much scatter.

He then introduced the idea of atomic number Z as the position of the element in the periodic table, and found that a new graph of the square root of the frequency against Z was a good straight line.

Thus Moseley's law was stated as

$$\nu \propto (Z-1)^2 \quad \text{for the } K_\alpha \text{ radiation.}$$

The concept of atomic number also helped to sort out the periodic table, for in certain cases the order of atomic number is different from the order of the atomic weights and the atomic number order fits the chemical properties much better.

Finally, in conjunction with Rutherford's work, Z was realized to be a measure of the nuclear charge on the atom.

51.3   Positive rays

If a discharge tube experiment is carried out in a tube with a perforated cathode, rays are observed on the far side to the anode. With some difficulty they can be deflected by a magnet, and shown to be positively charged.

These rays were thoroughly studied by J. J. Thomson, and in particular he measured the value of e/m for them.

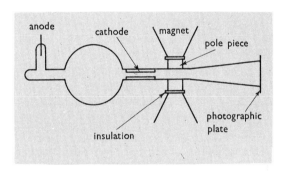

Fig. 325   J. J. Thomson's apparatus

Positive rays were produced by the discharge in the bulb from which most of the gas had been removed. Some passed through a fine hole in the cathode, which required water cooling because of the bombardment by many positive rays.

If no fields were applied, a spot was formed on a photographic plate. However, if parallel magnetic and electrostatic fields were applied by the insulated pole-pieces of the magnet, a series of parabola were observed on the photographic plate.

Let the positive rays have charge E (coulombs), mass M (kg) and velocity v (variable)

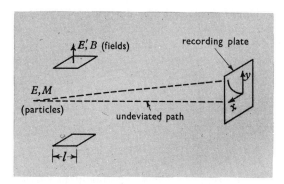

Fig. 326   The deflection of positive rays

(m sec^{-1}). Suppose they experience fields E' (volt m^{-1}) and B (tesla).

Then while they are in the fields of 'length' l they move with constant velocity in the direction perpendicular to the fields, as no force acts in this direction.

If this part of the motion takes time t

$$t = l/v$$

and the y deflection (due to the field E') is

$$y = \tfrac{1}{2}at^2$$

$$= \tfrac{1}{2}\frac{EE'}{M}\left(\frac{l}{v}\right)^2$$

where a = acceleration

and the x deflection (due to field B) is

$$x = \tfrac{1}{2}a't^2$$

$$= \tfrac{1}{2}\frac{BEv}{M}\left(\frac{l}{v}\right)^2$$

and eliminating v

$$x^2 = \frac{E}{M}\frac{B^2l^2}{2E'} \cdot y$$

which is the equation of a parabola. So all particles of the same E/M but with different velocities lie on one parabola. Now moving the photographic plate further back, merely brings in a geometrical constant depending on the shape and size of the apparatus.

By comparison of the parabolae the masses of different atoms may be compared, for the positive rays are found to be ionized atoms, basically of similar charge (there may be small multiples), but of varying mass.

In 1912 neon was used in the bulb of such an apparatus, and two distinct parabolae were found corresponding to masses of 20 and 22 units. The latter was much fainter, suggesting a smaller amount of the heavier variety, fitting in with the known chemical atomic weight of 20·2. Thus the idea of isotopes, atoms of the same chemical form but of differing weight, came into use.

Further development of these principles lead to mass spectrographs, capable of measuring the masses of atoms very much more exactly.

52   THE STRUCTURE OF THE NUCLEUS

52.1   The proton

When α-particles were allowed to strike hydrogen atoms in a vessel complete with zinc sulphide screen, scintillations can be observed when the screen is further from the α-source than the known range. These are due to the arrival of the hydrogen nucleus or proton at the screen after it has been struck by an α-particle, the electron associated with it having been left behind.

In 1919 Rutherford showed that many nuclei contain protons.

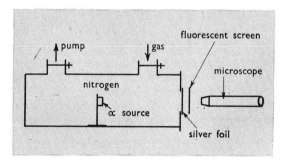

Fig. 327   The discovery of protons in the nitrogen nucleus

Air in the apparatus was replaced by nitrogen, which was subject to α-bombardment. The silver foil was sufficient to stop the α-particles, but scintillations, similar to those caused by protons, were observed. Rutherford was able to show that the particles causing the scintillations had similar properties to protons, sufficiently well to justify the idea that all nuclei contain protons.

52.2   The neutron

In 1930 it was found that the bombardment of beryllium with α-particles produced a very penetrating radiation of no charge. It was first suggested that it was very high energy γ-radiation.

In 1932 Curie-Joliot placed paraffin wax in the path of the penetrating radiation and found that protons were emitted. By calculating back he showed that if it were γ-radiation causing the emission, it had an improbably high energy.

Also in 1932 Chadwick further investigated the velocity of these protons by absorption measurements.

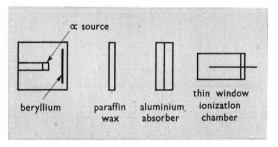

Fig. 328   The discovery of the neutron

His results were consistent with the idea that a new particle of zero charge, and mass similar to the proton were produced when the alpha particles bombarded the beryllium. This particle was called the neutron, and must be part of the nucleus.

Thus the nucleus was considered to be built up from protons and neutrons, together called nucleons.

This enables us to describe any nucleus in the following terms:

An element of atomic number Z and mass number A has Z protons and $(A - Z)$ neutrons in the nucleus. The mass number is the whole number nearest the atomic weight.

Any type of nucleus can be fully described by writing

$$_Z^A X$$

where A = mass number, that is the number of neutrons plus protons,

Z = atomic number, that is the number of protons,

X = corresponding chemical symbol.

Then a nuclear reaction can be represented symbolically.

e.g. The Rutherford experiment described above is represented by

$$_7^{14}N + _2^4He \rightarrow _8^{17}O + _1^1H$$

52.3   The effects of radioactive decay

(i) α-Emission. An alpha particle is the nucleus of the helium atom $_2^4He$ so its emission results in a drop in mass number of four and in atomic number of two

e.g. $_{88}^{226}Ra \rightarrow _{86}^{222}Rn + _2^4He$

(ii) β-Emission. A beta particle is an electron coming from the nucleus. It has negligible mass but unit negative charge, so its emission causes no change in mass number but an increase of one in atomic number.

e.g. $_{82}^{210}Pb \rightarrow _{83}^{210}Bi + _{-1}^0e$

(iii) γ-Emission. As gamma rays are electromagnetic they have neither mass nor charge, and therefore their emission causes no change in mass number or atomic number.

Where one nuclide (i.e. a particular type of nucleus) decays into another which is itself radioactive, a whole radioactive series may occur. Four distinct series occur among natural radioactive materials, namely, the uranium, actinium, thorium and neptunium series.

52.4   The structure of isotope nuclei

The isotopes of any element all have the same atomic number, but varying mass number due to variations in the number of neutrons in the nucleus.

e.g.
$_{10}^{20}Ne$ has 10 protons and 10 neutrons
$_{10}^{21}Ne$ has 10 protons and 11 neutrons
$_{10}^{22}Ne$ has 10 protons and 12 neutrons.

52.5   Isobars

These are different nuclides with the same mass number, e.g. $_6^{14}C$ and $_7^{14}N$ have the same mass number but different numbers of neutrons and protons.

52.6   Artificial radioactive nuclides

Whilst many radioactive nuclides are found to occur naturally, the total number now available is much greater owing to the artificial production of nuclides which do not occur naturally.

Artificial nuclides can be prepared by bombardment of other nuclides with various particles. These often need to be accelerated to very high speeds by one of the many large accelerating machines available.

More common is the use of neutrons in the atomic pile where the decay of $_{92}^{235}U$ produces the neutrons and a dynamic equilibrium can be maintained. The materials to be irradiated are placed in the pile and exposed to the neutrons for about six half-lives of the desired nuclide for maximum activity.

After a neutron has been captured other particles may be emitted

e.g. $_{17}^{35}Cl + _0^1n \rightarrow _{16}^{35}S + _1^1H$

or $_{17}^{35}Cl + _0^1n \rightarrow _{15}^{32}P + _2^4He$

and then chemical separation can be used.
N.B. The speed of neutrons has a great effect, the above equations are for fast neutrons, whereas for slow neutrons

$$_{17}^{35}Cl + _0^1n \rightarrow _{17}^{36}Cl + \gamma.$$

Radioactive isotopes are finding many uses in industry, research and medicine.

52.7   Examples of their use

A typical industrial use is in a thickness gauge. These can be of many kinds but one of the simpler is in the Paper industry. The β-radiation from thallium $^{204}_{81}$Tl is passed through the continuous stream of paper being produced and is attenuated according to the thickness of the paper. The resulting signal can be servo-connected to the machine to make it self-correcting.

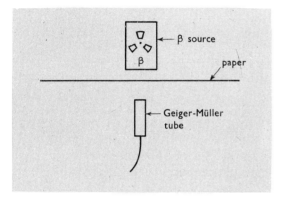

Fig. 329   A radioactive thickness gauge

A typical biological use would be to detect the distribution of trace elements in a leaf. A suitable radioactive nuclide is 'fed' to the plant and the leaf placed on a photographic plate. Blackening would indicate the presence of the radioactive material.

52.8   Mass and energy

Using the mass spectrograph, very precise values of nuclide masses can be obtained. The basic unit, the atomic mass unit (a.m.u.) which used to be defined as one-sixteenth of the mass of the oxygen atom $^{16}_{8}$O, but is now taken as one-twelfth of the mass of the carbon atom $^{12}_{6}$C, and is equal to $1 \cdot 660 \times 10^{-27}$ kg.

In 1905 Einstein showed that mass and energy are related by the equation

$$E = mc^2$$

where E = energy
m = mass
c = velocity of light

and so energy and mass can be interconverted.

The unit of energy often used in nuclear physics, the electron volt, is equal to the energy gained by an electron in moving through a potential difference of one volt, and can be equated to mass.

$$1 \text{ a.m.u.} = 931 \times 10^6 \text{ eV or } 931 \text{ MeV}$$

where 1 MeV = 1 million electron volts

52.9   Mass defect, binding energy

The mass of any nuclide nucleus differs from the total mass of its component nucleons:

e.g.
Helium nucleus has a mass of $4 \cdot 0028$ a.m.u.

a proton has a mass of $1 \cdot 0076$ a.m.u.

and

a neutron has a mass of $1 \cdot 009$ a.m.u.

$\therefore$

2 neutrons + 2 protons has a mass $4 \cdot 0332$ a.m.u.

and

the mass defect $= 4 \cdot 0332 - 4 \cdot 0028$
$= 0 \cdot 0304$ a.m.u.
$= 28 \cdot 3$ MeV

and in the latter form is called the binding energy of the nucleus, the energy necessary to pull the nucleons apart, and comes from their total mass, hence causing the mass defect.

The binding energy per nucleon is found to be generally constant at about 8 MeV, but lower at small atomic numbers.

52.10   Fission and fusion

The uranium nucleus on capture of a neutron splits into two comparable parts

$$^{235}_{92}U + ^{1}_{0}n \rightarrow ^{148}_{57}La + ^{85}_{35}Br + 3^{1}_{0}n$$

(etc.)

and masses

$$\underbrace{235\cdot1 + 1\cdot009}_{236\cdot1} \qquad \underbrace{148\cdot0 + 84\cdot9 + 3 \times 1\cdot009}_{235\cdot9}$$

so 0·2 a.m.u. or 186 MeV of energy is released, and the fission of 1 lb. of uranium released 10^7 kilowatt hours of energy.

Light nuclides can combine or fuse and again energy is released.

e.g.
$$^2_1H + {}^2_1H \rightarrow {}^3_2He + {}^1_0n$$

$$\underbrace{2\cdot015 + 2\cdot015}_{4\cdot030} \qquad \underbrace{3\cdot017 + 1\cdot009}_{4\cdot026}$$

so 0·004 a.m.u. of mass has been converted into 3·7 MeV of energy, but a very high temperature is needed for this reaction to occur.

53 THE FULL STRUCTURE

53.1 The whole atom

The 'picture' of the atom built up so far is of a composite central nucleus surrounded by electrons. The mass is concentrated in the nucleus and the overall charge is zero because the nucleus and the electrons carry equal and opposite charges.

However, there is a fine structure amongst the electrons, as these are grouped in shells or layers containing up to a maximum number in each shell.

Each shell and subshell is associated with a definite energy which is quantized.

53.2 Line spectra

When an atom gains extra energy, say by collision, an electron may become excited and move to a higher energy level. As this is unstable it falls back to the lower energy level emitting a photon in the process.

$$E_1 \quad E_2 = h\nu$$

53.3 The periodic table

The filling up of the successive shells with electrons as the atomic number increases gives rise to the explanation of the regularity of the periodic table.